T0219247

THE CULTURAL
CONTRADICTIONS
OF DEMOCRACY

The Cultural Contradictions Of Democracy

Political Thought since September 11

JOHN BRENKMAN

PRINCETON UNIVERSITY PRESS   *Princeton & Oxford*

Published by Princeton University Press, 41 William Street, Princeton, New Jersey 08540

In the United Kingdom: Princeton University Press, 3 Market Place, Woodstock, Oxfordshire OX20 1SY

Library of Congress Cataloging-in-Publication Data

Brenkman, John.
The cultural contradictions of democracy : political thought since September 11 / John Brenkman.
p.   cm.
Includes index.
ISBN: 978-0-691-17120-3   (alk. paper)
1. Political science—Philosophy.   2. Democracy.   I. Title.
JA71.B735 2007
320.01—dc22   2007007659

British Library Cataloging-in-Publication Data is available

This book has been composed in Adobe Garamond

Printed on acid-free paper. ∞

pup.princeton.edu

Printed in the United States of America

10 9 8 7 6 5 4 3 2 1

FOR HÉLÈNE

CONTENTS

INTRODUCTION: POLITICAL THOUGHT
IN THE FOG OF WAR   1

    War and Democracy   1
    Hobbes versus Kant?   4
    Leviathan   6
    The Neoconservative Illusion   9
    The Frailty of Human Affairs   12
    Crises of the Republic   14
    The Argument   19

SEIZED BY POWER   24

    Death and the Governor of Texas   24
    The New American Exceptionalism   28
    The Cold Warrior Myth   34
    Kant with Arendt   37
    Targeting Iraq   41
    Al Qaeda and Ultimate Ends   43
    A Grammar of Motives   46

THE IMAGINATION OF POWER   51

    State of Exception   51
    Arendt versus Agamben   55
    Schmitt and Hobbes   59
    Decision and Covenant   64
    The Ordeal of Universalism   71

SEPTEMBER 11 AND FABLES OF THE LEFT   78

    First Response   78
    Multilateral Ambivalence   81
    Terrorism as Symptom   84

Chomskian Certitudes 87

Hardt and Negri's *Empire* 94

The Multitude and Prophecy 98

IRAQ: DELIRIUM OF WAR, DELUSIONS OF PEACE 103

The Idealism of Means 103

The Idealism of Ends 106

Neither Left nor Right 110

The Atlantic Misalliance 117

Diplomatic Intrigues and Political Truths 122

Repudiations of the UN Left and Right 126

The Hobbesian Nightmare: Occupied Iraq 131

THE ORDEAL OF UNIVERSALISM 137

Democracy and War 137

Postnational Cosmopolitanism versus
    Liberal Nationalism? 141

Kant with Hobbes 144

Habermas's *Agon* with Schmitt 146

Hobbes with Kant 152

Europe, or, the Empire of Rights 157

Islam's Geo-Civil War 165

Global Neoliberal Religious Conservatism? 170

No Exit 177

CONCLUSION: PRELUDE TO THE UNKNOWN 182

Ideas and Errors 182

Arendt with Berlin 183

Liberty without Democracy versus Democracy
    without Liberty? 188

Democratic Striving and Sectarian Mobilization 191

Untimely Meditation 195

Index 201

I'm sentimental, if you know what I mean.

I love the country, but I can't stand the scene.

—Leonard Cohen, *Democracy*

THE CULTURAL

CONTRADICTIONS

OF DEMOCRACY

A dialogue, which is the highest form of communication
we know, is always a confrontation of irreducibly
different viewpoints.   —OCTAVIO PAZ

INTRODUCTION: POLITICAL
THOUGHT IN THE FOG OF WAR

War and Democracy

Since September 11, 2001, the fog of war has enveloped political
thought. Bright hopes of perpetual peace and prosperity collapsed in the
debris of the World Trade Center. The fog grew thicker with the invasion of Iraq in 2003, as the nations of the Atlantic alliance collided over
policy and principle, law and interests. By the time the postwar in Iraq
became a civil war and produced more casualties than the war, the dominoes that neoconservatives dreamed would democratize the Middle East
were falling helter-skelter into new uncertainties. On the other side of
the debate, the most dire antiwar prophecies seemed exaggerated if not
hollow, when Iraqis managed to hold elections for the first time in
decades. Neither the advocates nor the opponents of the war in Iraq had
any sure insights into the uses of war on behalf of democracy. When, by
the fall of 2006, the number of Iraqis being killed every month as a result of civil strife exceeded the number of Americans who had been
killed in the September 11 attacks that had supposedly justified the invasion of Iraq, American policy was losing its political as well as moral
bearings.

Combatants easily lose their sense of direction in the midst of battle,
confuse comrade and foe, mistake progress for setback and setback for
advantage, and retreat when on the verge of victory or hurl themselves
into certain devastation convinced of their invincibility. Nor does war
spare political thought from disorientation and uncertainty. Fundamental questions of war and democracy had scarcely begun to emerge

1

into public awareness after September 11 when they were swept up in the whirlwind of preemptive war in Iraq. Among those fundamental questions are:

What role do arms have in a democracy?

How does military power alter, as well as protect, the polity that uses it?

What is the possibility and even the meaning of the international rule of law?

Can force effectively spread liberty and democracy abroad?

The spectacularly successful interventions that overthrew the Taliban and then Saddam Hussein were quickly compromised by the occupations of Afghanistan and Iraq. A pattern emerged. *The United States overestimates the effectiveness of military might and underestimates the ordeal of democracy.* Responsibility for the appalling shortcomings of the postwar rebuilding of Afghanistan and Iraq falls squarely on the presidency of George W. Bush—but not exclusively. The crisis runs deeper than any one administration when the world's oldest democracy and sole superpower does not comprehend the wellsprings of democracy or understand the nature of might. Opponents of Bush, neoconservatives, and the Republican Party delude themselves when they are satisfied merely with opposition to administration policy. The fundamental questions of war and democracy are even more difficult to answer today than on September 11, 2001. The threat of terrorism has likely grown rather than shrunk since 2001, and the Middle East has been turned into a laboratory of democracy where any failed experiment risks producing civil war, dictatorship, theocracy, or worse.

Diplomatic and military decisions are prepared and justified by a discourse authored by many hands: historians and ideologues, politicians and journalists, scholars and pundits. A range of discourses, from think-tank manifestos to "Great Power" historiography, from secret reports to presidential speeches, from strategic studies to op-ed pieces, produces the intellectual—but also the symbolic—webbing of the decisions and actions taken in foreign affairs. Since September 11, this symbolic-conceptual webbing has teemed with terms like "lone superpower," "hyperpower," "liberal imperialism," and "progressive interventionism," and slogans like *war on terror* or *power-vs-weakness* and *Hobbesian-vs-Kantian.* All these slogans imply some understanding of power; they all imply, to draw on a distinction made by Hannah Arendt, some understanding of political *power* and military *might*. Metaphors as well as ideas are at work in foreign policy discourse, for not only are there conceptions of power—and various methods of analyzing, say, the relative power of states or calculating their interests—but there is also an *imagination* of power. When it comes to military force no one can truly know how its use will enhance or diminish the power of the state that wields it. Might exists primarily in potential, and therefore it always exists in the imagination.

In liberal democracies all newly elected leaders, all new heads of state, find themselves suddenly in possession of power. And, inversely, they find themselves possessed *by* power. There is an inescapable ambivalence in the enjoyment of power. Max Weber turned to a bodily image when he identified the greatest of the "inner enjoyments" of the vocation of politics: "the feeling of holding in one's hand a nerve fiber of historically important events."[1] To *enjoy power*—or to *be empowered*—is at the same time to be enjoyed *by* power. Having the means of coercion and violence in your grasp puts you in the grip of those very means of coercion and violence. That is why Weber, attentive to the insights of Nietzsche and riveted to the upheavals of war and revolution in his own time, insisted that the ethics of political life must include an awareness of politics' inherent potential for tragedy. Such an awareness alone tempers the intrinsic temptation to a kind of power beyond responsibility.

To become president of the United States in 2001, as happened to George W. Bush, was to find oneself suddenly in possession of power-beyond-responsibility, since the American body politic itself had been in the grip of power in excess of responsibility for a decade. The collapse of the Soviet Union left the United States with unmatched military might. It faced historic questions: *For what ends does the nation possess such means of violence? Should this nation, or any nation, have unmatched military might? What responsibilities attend overweening power?* The body politic never answered these questions or even seriously debated them. There were, of course, many discussions in the government, foreign policy think tanks, and opinion and policy journals, but the presidential

[1]Max Weber, "Politics as Vocation," in *From Max Weber: Essays in Sociology*, ed. and trans. H. H. Gerth and C. Wright Mills (New York: Oxford University Press, 1946), p. 115.

campaigns of 1992, 1996, and 2000 avoided the controversy altogether. It became commonplace to refer to the Gulf War as the watershed of the United States's emergence as sole superpower, and politicians and theorists coined such grandiose names for the new era of American supremacy as the New World Order, the end of history, *hyperpuissance,* Empire. None of the labels took into account the stark fact that the American body politic remained silent and indecisive.

September 11 changed all that, though not because a great national debate finally occurred. Such a debate did not occur. Rather, after the Taliban and al Qaeda were routed in Afghanistan, the Bush administration was emboldened to advance an answer to the historic post–Cold War questions. In the 2002 document called *The National Security Strategy of the United States,* President Bush called upon the United States to embrace the role of supreme global power. With this appeal came the declaration of America's unique right over other nations, a threefold right to preemptive war, the overthrow of regimes considered hostile, and immunity from treaties and constraints imposed on other nations.

Hobbes versus Kant?

The doctrine of unilateralism and preemption contributed perhaps more than anything else to the showdown between the United States and major European allies, especially France and Germany, in the build-up to the war in Iraq. The U.S.-European divergence seemed neatly to confirm Robert Kagan's diagnosis that "Americans are from Mars and Europeans from Venus." Kagan is a particularly pertinent reference point, since the vision put forth in the Bush doctrine finds its intellectual backing in that strand of contemporary conservative thought represented by Kagan and preoccupied with the theory of "great powers." In his book *Of Paradise and Power,* which grew out of the essay "Power and Weakness" that stirred considerable discussion in the United States and Europe in the summer of 2002, Kagan ostensibly attempts to explain the markedly different views of Americans and Europeans in foreign affairs, in particular, the American inclination to unilateralism and force and the European preference for internationalism and negotiation in responding to crises: "On the all-important question of power—the efficacy of power, the morality of power, the desirability of power—American and European perspectives are diverging. Europe . . . is entering a post-historical paradise of peace and relative prosperity, the realization of Kant's "Perpetual Peace." The United States, meanwhile, remains mired in history, exercising power in the anarchic Hobbesian world where international laws and rules are unreliable and where true peace and security and the defense and promotion of a liberal order still depend on the possession and use of military might."[2] Kagan makes many interesting arguments and observations on U.S.-European relations, all more or less debatable, but it is the axioms that frame his whole discussion which throw a light on the mentality of the Bush administration and its understanding of power and military force. The fundamental axiom is simply a tautology: weak is weak, strong is strong: "When the United States was weak [in the eighteenth and nineteenth centuries], it practiced the strategies of indirection, the strategies of weakness; now that the United States is powerful, it behaves as powerful nations do. When the European great powers were strong, they believed in strength and martial glory. Now, they see the world through the eyes of weaker powers. These very different points of view, weak versus strong, have naturally produced differing assessments of threats and of the proper means of addresing threats, and even differing calculations of interest."[3] The tautology radiates out into the entire essay as the words *nature, normal, perfectly normal, predictable, naturally* alight on every aspect of current American policy just to say that a powerful nation does as powerful nations do. Policy flows from might. This axiom, presented by Kagan in the flat, frankly amoral tones of the historian of "great powers," undergirds the moral hyperbole by which the politicians Bush and Cheney justify the ambitious designs of the new National Security Strategy.

Is is it really Kantians versus Hobbesians, Venus versus Mars? Both slogans are catchy, and Kant-versus-Hobbes has even caught on among serious political philosophers. But these oppositions do not hold up philosophically or politically. It is hard to imagine a thinker less venereal than Kant. And, just as strikingly, there is nothing at all martial about

---

[2] Robert Kagan, *Of Paradise and Power: America and Europe in the New World Order* (New York: Alfred A. Knopf, 2003), p. 3.

[3] Ibid., pp. 10–11.

Hobbes. For starters, Hobbes fled England during the Civil War and wrote *Leviathan* in Paris in order to exorcise his horror at the image of civil order breaking down. Such a breakdown exposes a "state of nature" in which every man would have the right to do whatever he deems necessary "for the preservation of his own Nature; that is to say, of his own life." A commonwealth, Hobbes reasoned, arises from the fear of death that pervades the hypothetical state of nature, that is, the "condition of Warre of every one against every one."[4]

Contrary to Kagan's depiction, Hobbes's view of international relations little resembles that of the neoconservatives. "An anarchic Hobbesian world where international laws and rules are unreliable" is in fact not Hobbesian at all. Hobbes considered states to be *less* prone to violence than individuals, primarily because a ruler has responsibility for the peace and security of his subjects, and thus was inclined in Hobbes's view to be measured in assessing when their benefit and general welfare was best served by war or conquest.

Leviathan

Although Hobbes's ideas do not truly support the neoconservatives' vision of power, there is surely something that prompts Kagan and others aptly to consider the Bush doctrine Hobbesian. What stirs their imagination is the image of the body politic as leviathan. This image evokes something that is conveyed less in Hobbes's own words than in the famous engraving that graced the first edition of *Leviathan*. The monarch, his body composed of nothing but the multitude of his subjects, a scepter in one hand and a sword in the other, wearing a crown, rises above the land, whose gently rolling hills and little villages are at once the realm his outstretched arms protect and a kind of robe spread out around him. The undulations of the hills also suggest the waves that are the element in which the biblical leviathan lives. For the leviathan, as Melville well knew, is a sea monster, the whale as grasped in the ancient Hebrew imagination. There, in the more symbolic stew of Hobbes's thought, where body politic, sea monster, and monarch blend together,

---

[4] Thomas Hobbes, *Leviathan*, ed. C. B. Macpherson (New York: Penguin, 1968), p. 189 [ch. 14].

the Anglo-Saxon political imagination has found, precisely, an *image* of state power. The state is a monster, the One preventing the anarchy of the Many, that floats unperturbed in the sea or sails the oceans of the world to keep its multitude at peace, prosperous, and safe. That Britain—the island commonwealth, the commonwealth as island—would imagine the body politic a great sea monster makes nearly immediate sense. Hobbes has not, however, enjoyed extensive influence on American political thought. The American imagination of power has historically been more isolationist (the stay-at-home leviathan of the Monroe Doctrine) and territorial (the land-bound behemoth of Manifest Destiny). So how has the Hobbesian image come to fit America?

In a news conference in April 2004, as Iraq was being shaken by the simultaneous insurgencies of Sunnis in Falluja and militia loyal to Moktada al-Sadr in the Shiite South, and as the 9–11 Commission was probing how much warning the administration had prior to the al Qaeda attacks, President Bush uttered, as though revealing something for the first time, an astonishing anachronism. "We can no longer hope," he declared, "that oceans protect us from harm." Americans have, of course, been quite aware that oceans do not protect us from harm ever since 1957, when the Soviets sent the first Sputnik into space and raised the specter of intercontinental ballistic missles raining nuclear warheads on American cities. Bush's own chief preoccupation in defense matters *before* September 11 had been the renewal of Star Wars, the missile shield project that was predicated on just such vulnerability across oceans. Not only had Americans known for nearly fifty years that the Atlantic and Pacific afforded no guarantee against attack, but the September 11 attacks themselves, though they were an unprecedented assault by foreigners against American civilians at home, did not originate from abroad: the planes all took off on American soil.

Was Bush's anachronism then simply a historical lapsus? Not at all. For what it did, like so many other carefully crafted misstatements and innuendoes of that moment, was to associate the September 11 attacks carried out by al Qaeda with the presumed weapons of mass destruction in the hands of the United States's self-styled enemy Saddam Hussein. The anachronistic image of the protective oceans created a link between September 11 (attack on U.S. soil) and Saddam Hussein (weapons of mass destruction). According to polls, by the time the war in Iraq began

45 percent of Americans believed Saddam Hussein had planned September 11, just as many believed that his missiles with their range of a few hundred kilometers could reach the United States.

The war on terrorism and the war in Iraq had nothing to do with one another. We are engaged in the first because we have to be; we engaged in the second because the Bush administration wanted to, and could. They *could* because they took office in possession of unmatched military might. The attack on the Twin Towers and the Pentagon showed our vulnerability; the invasion of Iraq was meant to show our superior strength. The illusion of invulnerability that had shattered on September 11 was quickly transformed into an illusion of insuperability in preparation for war in Iraq. Put the two scenarios together and America becomes Hobbesian: the American body politic, no longer unperturbed floating in its oceans, gets transformed in the political imagination into the unvanquishable monster sailing the seas of the world. Shock and awe in Iraq answered, in symbol and fantasy, the shock and awe of September 11. Americans were called upon to see America the global power in leviathan imagery: "When he raiseth up himself, the mighty are afraid. . . . The sword of him that layeth at him cannot hold: the spear, the dart, nor the habergon. . . . Upon earth there is not his like, who is made without fear" (Job 42:25–33).

The new national leviathan, however, hardly seems like one "who is made without fear." After September 11, political leaders and the media cultivated and relentlessly fertilized an all-pervasive fear within the American body politic. The sense of vulnerability and dread often seemed far more intense in the country at large than among the New Yorkers who had actually witnessed the destruction in their own city. Small towns in rural areas believed themselves to be targets in imminent danger when they learned that the federal government was furnishing local authorities with gas masks or special medical supplies. Unlike Roosevelt declaring that the only thing to fear was fear itself or Churchill exhorting Londoners to persevere in the face of the bombings that nightly ravaged their city, President Bush used September 11 to stoke citizens' fear for their lives. The most striking characteristic of the national mood between September 11 and the invasion of Iraq was how this excessive fear was strangely combined with inordinate confidence that military might could guarantee international and national security. Even as the administration purveyed fear and vulnerability, it promised that the army would topple dangerous tyrants armed with weapons of mass destruction and create new democracies. Therein is the mainspring of the administration's pseudo-Hobbesian vision and rhetoric. The national leviathan fuses an all-pervasive amorphous fear of death and an overweening faith in military power.

What are the origins of this combination of fear and hubris? Why was this combination so effective in cementing public opinion in support of the war? The answer undoubtedly has a lot to do with the post-Vietnam demise of a military of citizen-soldiers. Having first professionalized and now increasingly privatized the nation's armed forces, the Pentagon has so separated citizens from soldiers that citizens are free to indulge in debilitating fear while fully expecting their soldiers to bravely conquer all. Another factor is the tendency within contemporary culture to privilege the figure of the *victim*. Symbolizing oneself as a victim lends an aura of moral rightness, a privileged viewpoint on what is just and unjust—and considerable entitlement and ultimately a certain license. The shock of September 11 was from the beginning nurtured into this dangerous mix of victimage, righteousness, and license. The leviathan motif expresses well the post–September 11 fear. This cultivated fear, this exaggerated yet real fear of death, has been integrated into the symbolic-discursive webbing of foreign policy. A wounded, half-blind leviathan thrashing about in geopolitical seas does not see the fragility of democracy at home and the difficulty of inaugurating it abroad.

The Neoconservative Illusion

The neoconservatives confuse power and might. Hannah Arendt makes the distinction in the sharpest of terms. *Power* "corresponds to the human ability not just to act but to act in concert." It is utterly distinct from might and violence. Might is the capacity for violence, and violence itself marks the breakdown of power, whether within the polity (where "violence functions as the last resort of power against criminals or rebels—that is, against single individuals who, as it were, refuse to be overpowered by the consensus of the majority") or externally in war, where in fact strength or might does not in itself secure power: "as for actual warfare, we have seen in Vietnam how an enormous superiority in the means of violence can become helpless if confronted with an ill-equipped but well-organized opponent who is much more powerful."[5]

From the Arendtian standpoint, all the tautologies of the neoconservative hawks exemplified by Robert Kagan's writings are false. Policy does not flow naturally, normally, predictably, from might. The nature of the polity is not immune to the effects of wielding the available means of violence. The unilateral exercise of force does not necessarily defend and protect the polity. That the colossal failure of America's anticommunist foreign policy in Vietnam, which stimulated the creation of neoconservatism in the first place, might turn out to be the very model of their own deepest errors is more than an exquisite irony. It risks becoming a historic tragedy. The consequences of the neoconservatives' conflation of power and might are seen not simply in theory but in practice. The crisis-ridden occupation of Iraq exposed the fundamental flaws in neoconservative thought, especially its Great Power discourse. The neoconservative hawks seem to have seriously believed that once American military might overthrew Iraqi tyranny, Iraqi democracy would spontaneously flower in its place; meanwhile, the more hard-nosed members of the administration assumed, just as naïvely though callously, that a strongman allied with the West could always step in to impose order if need be. The two attitudes resulted in gross negligence when it came to planning the occupation. Defense Department planning, directed by Donald Rumsfeld's undersecretary Douglas Feith, thoroughly ignored the need for civil order. The American occupation created the conditions for Hobbesian anarchy *inside Iraq*. Iraq turned into a breeding ground for Islamic terrorism because it lacked civil order and was awash in weapons and rife with contending factions. The Coalition Provisional Authority headed by L. Paul Bremer III replaced Saddam Hussein's rogue state with an insecure failed state.

Iraqis owe their tyrant's overthrow to the Americans, but those Iraqis who heroically risked life and limb in an effort to salvage a democratic path owe Americans little else. For months on end after the invasion, Iraqis were driven to arm themselves and ally, often unwillingly, with

---

[5]Hannah Arendt, "On Violence," in *Crises of the Republic* (New York: Harcourt Brace & Company, 1972), pp. 143, 150.

militias or mullahs in the face of absolute uncertainty over their own safety. They felt the naked insecurity of the state of nature. The millions who exercised their right to vote under threat of death added a new and poignant symbol to democratic imagery—and, indeed, to the Hobbesian image-bank too: their ink-stained fingers are forever an emblem of individuals mastering the fear of death in order to create a civil order that might protect them all. What these Iraqis have Rumsfeld and Bush, Feith and Bremer, to thank for is leaving them unprotected. Fear and death swirled around their first election and continued to plague their civil covenant.

The neoconservative understanding of power also fails to comprehend that the might which a democracy possesses and uses can flow back—or blow back—to alter the very nature of its own democratic institutions. The neoconservatives conceive of the nation's liberal order and rule of law as an unalterable internal feature of American democracy, while armed force is simply the means of protecting the nation against external threats. The illusion that American democracy itself cannot be harmed by the might it wields abroad has proved costly. Consenting to war is the most serious decision that citizens have to make. When the reasons by which citizens are persuaded to give their consent are, as with Iraq, erroneous (weapons of mass destruction) and deceptive (Iraqi ties to al Qaeda), the very fabric of democratic deliberation is damaged. The insistence that the errors and deceptions did not ultimately matter, since the result was good, even further undermined the very principle of accountability; a re-elected George Bush retained or promoted all the officials most responsible for the errors and deceptions. The posture that right-or-wrong we are right weakened the value put on truth in public affairs; the administration successfully fostered, with the help of the Fox-led media, a hatred of dissent, all the more intensely when the dissenters accurately exposed administration falsehoods. The willful distortion of international and domestic law through *enemy combatants, detainees, homeland security, coercive interrogations, extraordinary rendition* eroded civil liberties and the rule of law at home just as it severely damaged the United States's moral standing abroad. And the transformation of the border into a digital fortress that indiscriminately makes every visitor and immigrant a suspect has estranged the very foreigners most likely to value the United States as a symbol and example of a democratic way of life. So many essential features of American democracy—open deliberative politics, a public sphere valuing debate and truthfulness, due process, the furnishing of a symbolic and real haven of freedom for people around the globe—have been strained and damaged by the use of force in Iraq.

The reassuring view that arms are merely the instrument of a secure democracy in an insecure world is the fateful illusion at the heart of American policy. This illusion has its pseudo-Hobbesian variation in Kagan's equation of political power and military might in the amoral language of Great Power discourse, and it has its romantic-messianic variation in Bush's rhetoric of the axis of evil, the forward strategy of freedom, the end of evil, and so on. In calling this vision of foreign affairs an illusion, I have in mind the sense that Freud gave to the term in *The Future of an Illusion,* namely, an account of reality that coincides with what one wishes reality to be. An illusion in this sense is not necessarily demonstrably false, but wherever reality seems to coincide with one's wishes, doubt and skepticism ought to precede action, especially when the action is making war.

The Frailty of Human Affairs

It is almost impossible for Americans to entertain the idea that their democracy might be fragile. The political system after all has endured for 220 years under the same Constitution. The duration perhaps fosters a false sense of continuity, since there may well be as many discontinuities as continuities between the United States of 1787 and today. Moreover, American democracy has been dramatically depleted or renewed at crucial moments in its history. The 1850s were the darkest decade in America's political life; the crisis that was leading to the Civil War revealed that neither of the great strands of American democratic values, neither liberalism nor republicanism, could light a path to the end of slavery. The 1960s were a moment of extraordinary democratic renewal, though the conservative opinion on the ascendancy today looks back and sees cultural and moral decline. The civil rights movement overturned apartheid in America by using civil disobedience to extend fundamental rights, including voting rights, to blacks, and the antiwar movement initiated another innovation in citizenship as young people organized to protest the purposes and effects of a war they were called upon to fight in; these movements in turn inspired a new wave of feminism and the creation of movements for gay rights. Democratic renewal lay in the innovation of rights and invention of freedoms. The upheavals of the 1850s and the 1960s are a reminder that even the most durable democracy undergoes erosions and needs reinaugurations.

Hannah Arendt relates the inherent fragility of democracy to the frailty of human affairs, but she also ties this frailty and fragility to the wellsprings of human creativity and political inventiveness. The Greeks founded their *polis,* she argues, in response to the frailty of what they most valued in their pre-polis experience for "mak[ing] it worthwhile for men to live together (*syzen*), namely, the 'sharing of words and deeds' [Aristotle]." The polis was "to multiply the chances for everybody to distinguish himself, to show in deed and word who he was in his unique distinctness." And secondly it was "to offer a remedy for the futility of action and speech; for the chances that a deed deserving fame would not be forgotten, that it actually would become immortal, were not very good." The polis was the boundaried, organized, sheltering space where individuals' words and deeds could appear and endure. The bid for immortality paradoxically rendered the polis itself mortal: "One, if not the chief, reason for the incredible development of gift and genius in Athens, *as well as the hardly less surprising swift decline of the city-state,* was precisely that from beginning to end its foremost aim was to make the extraordinary an ordinary occurrence of everyday life."[6]

The Athenian paradox illuminates a dynamic of the political realm in general, namely, that its power and its fragility have the same source. The human gathering that lets speech and action *appear,* that is, manifest themselves *publicly,* to those that gather and participate defines power in the Arendtian sense: "Power is what keeps the public realm, the potential space of appearance between acting and speaking men, in existence." Forms of government in her definition are "the various forms in which the public realm can be organized." A political community depends on "the unreliable and only temporary agreement of many wills and intentions." Therein lies at once the power and the

---

[6]Hannah Arendt, *The Human Condition* (Chicago: University of Chicago Press, 1958), pp. 196–97. My italics.

fragility of political community: "[Power's] only limitation is the existence of other people, but this limitation is not accidental, because human power corresponds to the condition of plurality to begin with. For the same reason, power can be divided without decreasing it, and the interplay of powers with their checks and balances is even liable to generate more power, so long, at least, as the interplay is alive and has not resulted in a stalemate."[7] If a body politic endures, it is not because there is anything unalterable in its institutions and values, but because this pluralistic agreement is continually replenished, renewed, reinaugurated.

Conversely, the body politic becomes more fragile with every separation of decision and participation, every undoing of checks and balances, and every divergence of word and deed. These temptations are all the more insidious because they so easily mask themselves as ways of enhancing power, especially for a political communty in the thrall of fear and the fog of war—as when the president usurps the judiciary and its principles to better "protect" the American people, or when the Congress abdicates its role in determining the legitimate cause for war and scrutinizing the conduct of war, or when the press lets itself be embedded in the war machine, or when the Supreme Court refuses to shield reporters from the government and thereby shields government from reporters. Arendt once more: "Power is actualized only where word and deed have not parted company, where words are not empty and deeds not brutal, where words are not used to veil intentions but to disclose realities, and deeds are not used to violate and destroy but to establish relations and create new realities."[8]

Crises of the Republic

Mindful of the extraordinary legacy of Arendt's political thought, I hope to preserve something of her spirit of inquiry and interpretation in the following chapters. Trying to reflect on the crises of the republic *in the midst of* those crises is an abrupt reminder that political thought never enjoys a firm grounding. Such intellectual ideals as the unity of

[7]Ibid., pp. 200–201.
[8]Ibid., p. 200.

theory and practice, the proceduralist commitment to philosophically justified norms, or the pragmatist belief in shared values and tested methods all prove rather feeble when it comes to understanding a democracy at war. The difficulties are all the more acute because the crises of the American republic since September 11 have disguised themselves by masquerading as the self-assured, democracy-protecting, faith-soaked assertion of national strength. The temptations to deception and self-deception are always great in the midst of war, and in some fundamental way they are intrinsic to political life. The public realm, in Arendt's account, is where individuals "disclose themselves as subjects, as distinct and unique persons" through "their deeds and words." It is a space of appearance but not transparence, of self-revealing but not necessarily self-understanding: "Although nobody knows whom he reveals when he discloses himself in deed or word, he must be willing to risk disclosure. . . . It is more than likely that the 'who,' which appears so clearly and unmistakably to others, remains hidden from the person, like the *daimon* in Greek religion which accompanies each man throughout his life, always looking over his shoulder from behind and thus visible only to those he encounters."[9]

While Arendt renewed the import of ancient democracy for modern politics, no one revealed the ethical paradoxes of modern politics more penetratingly than Max Weber. Looking at the professional politician's psychological and ethical existence, Weber identifies the enjoyments of this vocation in the "feeling of power" and "knowledge of influencing men." The aptitude for such a career amalgamates "passion" for a cause, "a feeling of responsibility" for the consequences of decisions and actions, and "a sense of proportion." Passion, responsibility, and proportion are, however, perpetually vulnerable to two essential instruments of political power: *pretending* and *violence*. Since every politician "works with the striving for power as an unavoidable means," he "is constantly in danger of becoming an actor as well as taking lightly the responsibility for the outcome of his actions and of being concerned merely with the 'impression' he makes." The politician is, moreover, drawn into a realm of recurrent ethical paradox, since at bottom "politics operates with very special means, namely, power backed up by *violence*." Whoever

[9]Ibid., pp. 183, 181–82.

holds or vies for political power "lets himself in for the diabolical forces lurking in all violence."[10]

Max Weber and Hannah Arendt are largely irreconcilable thinkers. The sociologist of modernity and the political philosopher of the ancient polis diverge sharply in their understanding of power and violence, and their respective projects are more dissonant than consonant. And yet Weber's thoughts on the vocation and ethics of modern politics and Arendt's on the ancient resources of the modern polis create a productive strife, a meaningful dissonance, for an understanding of democracy at war today. Drawing on the contradictory traditions of democratic thought is in fact vital to political thinking. The plurality and debate so essential to democracy render disputes over the meaning of democracy a feature of democracy itself. Just as I draw on the contradictory ideas of politics and power developed by Weber and Arendt, so too I will adhere to the contradictory ideas of freedom found in Arendt and Isaiah Berlin. Arendt's supreme value is self-rule through participation in a body politic, that is, in government; Berlin's supreme value is the individual's freedom from the constraints of others and especially government, that is, the body politic. Therein lies the permanent clash between the civic and the liberal dimensions of modern democracy. As American democracy plunged into war, it fused its justifications for self-defense with the ambition to overthrow tyrannies and spread democracy. Its understanding of self-rule, liberty, and power is at stake in these acts and justifications, just as the acts and justifications are at stake in the understanding of self-rule, liberty, and power.

What then is the vantage point from which I undertake the work that follows? I consider the war against terrorism a necessity in which the democratic world will be engaged for several years. The overthrow of the Taliban in Afghanistan was in my view a justified and measured response to September 11. Moreover, I do not think that the invasion of Iraq to overthrow Saddam Hussein was immoral or illegal. It was, however, ill-advised and ill-conceived, and the failure to secure civil order in Iraq was unconscionable. American arrogance was matched by American ignorance. However, "withdrawal," "exit," and "bring the troops home" are the empty slogans of an antiwar movement without a

[10]Weber, "Politics as Vocation," pp. 114, 116, 119, 125–26.

vision of what role the United States needs to play in the Middle East. American policy under Bush has probably spread terrorism more than democracy and intensified rather than diminished international insecurity. By the same token, the status quo ante in the Middle East was untenable, and Islam's geo-civil war could not, and cannot, simply be "contained" on the Cold War model of containing communism.

I also start with the premise that while the flaws in American policy are decidedly the responsibility of the Bush administration, their source lies deep within the body politic itself. Overestimating the effectiveness of military force, Americans dangerously disregard the fragility of democracy at home and the ordeal of democracy abroad. Such disregard for the conditions of democratic life compounds the bad judgments made in Iraq and elsewhere. The next administration can easily temper Bush's brashness and de-emphasize unilateralism, and it is certainly unlikely to look forward to invading and occupying any other country in the axis of evil. The next president might even boldly shut down Guantánamo and undo all the executive orders that have permitted false imprisonment, torture, and "rendition." But will a new president and administration know how to pursue the war against terrorism without exacerbating the breeding grounds and motives of terrorism? Will they know how to lead the world's democracies' response to Islam's geo-civil war without ultimately simply aligning with the friendliest autocrats capable of repressing the restless Muslim masses? Will they know how to repair the American commitment to the rule of law and civil liberty enough to rescue America's standing as an inspiration for those who aspire to freedom and self-rule?

In taking democracy at war as my theme, I am looking to understand how political thought faces the difficult questions prompted by September 11 and the war in Iraq. These two unprecedented events have forced new avenues of inquiry and interpretation. Al Qaeda's September 11 attacks were an act of war by a nonstate actor, whereas political thought and foreign policy have long understood war as armed conflict *between* states or, in civil war, *within* a state. Unsettled questions abound concerning sovereignty, the meaning of the "war against terrorism," and its sources of legitimacy. The invasion and occupation of Iraq were no less unprecedented than September 11. The United States under George W. Bush not only justified the invasion of another country on the grounds of overthrowing tyranny and spreading democracy, but did so under the banner of unilateralism and preemption and in effect as a display of its own global supremacy. This step onto uncertain terrain was taken without public deliberation and decision, for even though the invasion itself had overwhelming public support, the reasons for going to war in Iraq were to overcome the impending if not imminent danger of its weapons of mass destruction and to sever its ties with Islamic terrorism. There were no weapons of mass destruction and no link existed between Iraq and Islamic terrorism until the American invasion and occupation *created* one. Whether the overthrow of tyranny and inauguration of democracy were the real motive of the invasion all along or the rationale retrofitted to an intervention that lost its original justification or, as is most likely, some mix of the two, the United States has been firmly set on a course in which the spreading of democracy is now the principal justification of its foreign policy, and the precedent of unilateral preemptive intervention remains for the moment unchallenged.

This situation poses a major question with which political thought is now grappling: Is there a vision of foreign policy uniquely suited to democracy? I myself am extremely skeptical—and often alarmed—when it comes to various efforts to claim a kind of organic connection between democracy at home and policy abroad. The post–Cold War swing of the pendulum away from *Realpolitik* as practiced by Nixon and Kissinger was undoubtedly an advance; their policies were often diabolic in outcome (as in Cambodia) or in intention (as in Chile), and the end of the Cold War certainly opened the possibility of a more principled adherence to political ideals. Nevertheless, I am going to question at various points the different visions of foreign policy as an organic extension of democracy. I have already begun to criticize how the neoconservatives' version of American exceptionalism promotes military supremacy as the natural way for a democracy to sustain and protect its liberal order at home. More idealistic visions of democracy's rightful recourse to violence as a means of spreading democracy have also been articulated. Paul Berman, for example, looks to weld liberal idealism to American military supremacy and make the elimination of tyranny the guiding principle of foreign affairs. Against neoconservative and liberal efforts to synthesize military might and democratic ideals, Jürgen Habermas advances the idea of cosmopolitanism. He envisions another kind of organic relation between democracy and foreign affairs, arguing that the rule-governed, deliberative democratic state should be the model for the international order just as it has been for the transnational order of the European Union. Where Robert Kagan foresees the American leviathan endlessly lashing out to destroy enemies of its democratic commonwealth, Habermas advocates a postnational order whose ultimate goal would be to transform action against terrorists, rogue states, and criminal cartels into the policing of a "global domestic policy."

Democracy at war stirs turmoil in political thought because the very aim of the internal workings of a democracy is to eliminate or transcend the use of force. Its institutions sustain that "unreliable and only temporary agreement of many wills and intentions" in which the democratic mentality delights. Neither military action abroad nor national-security measures at home foster such delight. On the contrary, military action stirs profound uncertainty, and security measures introduce the risk that the steps taken to protect democracy will dessicate its very values and institutions. By the same token, any realistic assessment of the threats and dangers posed by terrorism, rogue states, and weapons of mass destruction confirms that the security of democratic nations requires concerted action on the international scene backed by the capacity and willingness to use force. It all depends on judgment, and one role that political thought ought to play in the political life of a democracy is to broaden and enrich the capacity for judgment.

The Argument

My intention in the course of writing this book has been to probe the drama of political thought in the face of the war against terrorism, the overthrow of Saddam Hussein, and occupation of Iraq. In each chapter, the uncertain events of contemporary history are measured against, and are used to measure, the ideas that animate democratic traditions and political debate:

"Seized by Power": From his days as governor in Texas to his role as commander in chief, George W. Bush embraced politics as vocation by eschewing rather than assuming responsibility. The ideological architects of his foreign policy were likewise seduced by the power-beyond-responsibility of American military might. As the United States has attempted to confront radical Islamism's "ethic of ultimate ends" under the aegis of neoconservative ideas and democratic messianism, it has put its own democratic "ethic of responsibility" at risk.

"The Imagination of Power": The view that American global power exemplifies the "state of exception" as developed in the ideas of Carl Schmitt and Giorgio Agamben has been taken up with renewed intensity after the scandal of Abu Ghraib and Guantánamo. I argue that the conception of sovereignty that emerges from Agamben's appropriation of the Nazi jurist's theory of power is a deceptively appealing criticism of the modern state and a symptom of the malaise of contemporary "radical" political thought.

"September 11 and Fables of the Left": Beyond the ill-tempered response of some leftists immediately after the terrorist attacks against the World Trade Center and the Pentagon, the unprecedented confrontation between Western democracy and Islamic radicalism has brought out how thoroughly the political judgment and imagination of the so-called Left is limited by its underlying sensibility now that the world is no longer defined by the polarities of the Cold War. The faultlines of this sensibility become discernible in Noam Chomsky's moral absolutism and Michael Hardt and Antonio Negri's prophetic lyricism.

"Iraq: Delirium of War, Delusions of Peace": The claims of American unilateralism and the invasion of Iraq unleashed intellectual and diplomatic disputes that suggest that no simple dichotomy of Left and Right explains international politics. Leftists are split on questions of values as well as strategy. Paul Berman's defense of the war in Iraq evoked an idealism of ends (and the supreme value of freedom from tyranny), while Jürgen Habermas's opposition to the war drew on an idealism of means (and the supreme value of the international rule of law). The very idea of an international rule of law in the framework of the United Nations was, by the same token, belittled from the neoconservative perspective of Michael J. Glennon and the neo-Marxist perspective of Perry Anderson. On the diplomatic scene, the Atlantic alliance was shaken by the dispute that pitted Germany and France against the United States and Britain. In scrutinizing this dispute, it becomes apparent that it derived from a complex weave of principles and interests, not a clearcut division between Kantian peacemakers and Hobbesian warriors. Moreover, even as the justifications for the war were soon exposed as unfounded and erroneous, many of the premises of peace—including the effectiveness of sanctions—turned out to have strengthened rather than challenged Saddam Hussein's tyranny.

"The Ordeal of Universalism": While the neoconservatives' freedom-versus-tyranny theme and Habermas's postnational cosmpolititanism have the look of a sharp Right-Left dichotomy, neither adequately grasps the international conflicts it is supposed to address. Islam, which is often said to be engaged in a civil war, is embroiled rather in a geopolitical civil war. American policy ran aground in Iraq because it misunderstood this geo-civil war and then exacerbated it by neglecting to secure the country after the invasion. The grand project of overthrowing a tyrant in order to initiate democracies throughout the Muslim world yielded, instead, a civil war within Iraq itself that drew terrorists from all over the world, strengthened the radical regime in Iran, and trained suicide bombers to be sent back to London. As alternatives to the Bush doctrine are sought by Western thinkers and politicians, the interpretation of Europe's postwar project has become crucial. Is it the delusional Kantian paradise caricatured by Kagan or the beginnings of the "global domestic policy" idealized by Habermas? Or is it, as I argue, an innovative political body—an Empire of Rights—that is still reluctant to embrace its own transformative ambitions? In light of Islam's geo-civil war and Europe's halting project of extending democracy and capitalism, I challenge Walter Russell Mead's vision of a global neoliberal religious conservatism that would unite the capitalist West and conservative Islam.

"Prelude to the Unknown": Admittedly, this is an equivocal title for a conclusion, but it is true I think to the fact that as American ambitions in Iraq have faltered, the global situation of terrorism, geo-civil war, and tyranny is more desperately than ever in need of vigorous responses. Have the failings in Iraq and elsewhere squandered the very idea that freedom and democracy be the guideposts of American foreign policy? Or, conversely, might the rich tradition of ideas of self-rule (Arendt's positive liberty) and individual freedom (Berlin's negative liberty), both of which affirm human plurality, serve as a language of political criticism for assessing the failures and dilemmas of American policy itself?

Threaded throughout the book is a reflection on two political thinkers, Hobbes and Kant, whose ideas are as irreconcilable as those of Arendt, Weber, and Berlin. The evocation of Hobbes and Kant in debates over American foreign policy has simplified and warped the relation between their respective ideas; irreconcilable though they are, Hobbes and Kant do not conform to the current caricatures, and their relevance for the dilemmas and problems of the present can be startlingly unexpected, when it is recognized that Hobbes's thought hinges just as much on the body politic's founding covenant as on the war of all against all, and that Kant puts the darkness of human nature at the very heart of his reflection on perpetual peace.

The idealistic fervor and self-confident messianism that have infused American declarations of an armed crusade for democracy betray a dangerous disregard for the prospect of tragedy that Weber said inevitably accompanies politics when it turns to violence as a means for achieving its ends. The denial of tragedy amounts to a denial of responsibility. Not just moral responsibility, but ultimately political responsibility. Consider the silence regarding civilian casualties in Iraq. The United States made no effort to estimate civilian deaths, or even to assist Iraqis in accounting for their dead. Neither the press nor the Democrats insisted that an invading force and occupying power had such an obligation and duty. The gross negligence should arouse the intensest moral outrage, but also patriotic outrage. Our country has gone to war not wanting to know what it does. From the day of the invasion right through the destruction and depopulating of Falluja and beyond, America's leaders, its press, its representatives, its public—in short, ourselves—have indulged in this cowardly *wanting-not-to-know*. The silence is a form of lying. But the deception may be nearer self-deception than deceit. For it is well-known in much of the world that America does not have the courage to take a real look at the destructiveness of its own acts. The country's prevaricating silence on the tens of thousands of Iraqis killed is all the starker when juxtaposed to the exaggerated and vociferous fear of death in the United States since September 11. There is no reliable calculus for measuring what damage this kind of irresponsibility does to American interests abroad and the spirit of American democracy at home.

Political decisions and military actions are often thoroughly justified even though they violate valid moral standards. This is why, as Weber argued, politicians and statesmen must take responsibility for the foreseeable consequences of their decisions. A body politic incapable of accepting responsibility for politically justified but morally wrong acts cannot long operate effectively on the stage of world affairs. A body politic that *wants not to know* the harm it does loses its capacity to judge how its use of force accords with its political goals. How long before such refusal of responsibility and loss of judgment destroy the legitimacy of the foreign policy itself? I do not pose this question in order to launch into a jeremiad. Nor can the question be answered empirically. A body politic seldom has the benefit of reliable warning lights. A different metaphor is perhaps more apt. The body politic's institutions and, in Montesquieu's phrase, the spirit of its laws are its skeleton. Silence, lying, Orwellian misnaming, hypocrisy and hubris, negligence and bravado, wanting-not-to-know: these are all processes that *resorb* a democracy, ravaging the bones at a rate that is unfelt and unseen.

He who seeks the salvation of the soul, of his own or of others, should not seek it along the avenue of politics.   —MAX WEBER

SEIZED BY POWER

Death and the Governor of Texas

What turn did the strife between *ultimate ends* and *responsibility* take on September 11, 2001? What has been the fate of *isolationism* and *interventionism?*

The death of nearly three thousand people on American soil destroyed the immunity from the civilian horrors of war that had long underpinned the nation's isolationism. The fact that the terrorists had come from a global network with outposts in England, France, and Germany, recruits from Egypt, Lebanon, and Saudi Arabia, financial resources in Saudi Arabia, and training in Afghanistan and Florida, made unilateral intervention nonsensical. As for the perpetual conflict in politics between ultimate ends and responsibility, American democracy confronted a new test of the ethic of responsibility. It faced a stateless, international enemy committed to violence in the pursuit of ultimate ends, indeed to violence as an ultimate end.

The burden of comprehending the link between ultimate ends and violence in Islamic fundamentalism, of reconceiving foreign affairs beyond isolationism and interventionism, and of forging a new ethic of responsibility fell on the unlikely shoulders of George W. Bush. I don't say this mockingly (Bush's easily satirized anti-intellectualism and malapropisms caused many to underestimate his aptitude for political leadership), and I don't say it to assail his indifference toward foreign affairs on taking office (he was hardly the first American president to lack diplomatic experience). It is a question, rather, of George W. Bush's aptitude for judgment in the wake of September 11, of his ability to reshape the ethic of democratic responsibility in the midst of national agony and international crisis. The politician's subjectivity is truly revealed, as Max Weber saw, when the ethic of responsibility is doubly tested by *violence* as a means in politics and by the *inner enjoyments* of power, which motivate the political vocation as such. The Weberian categories call for an analysis not so much of the politician's psyche or intellect as the shape of his or her moral and ideological *daimon.*

The decision to undertake military action against al Qaeda and the Taliban in Afghanistan was in my view responsible, measured, and justified. But George W. Bush's daimon already began to show on the eve of the offensive in Afghanistan. Anticipating the arrest of al Qaeda members in the United States as well as the capture of others in Afghanistan, including perhaps Osama bin Laden, he gave extraordinary powers to himself and the military for dealing with presumed terrorists apprehended abroad or at home, including a plan for secret trials conducted at sea by military tribunals empowered to impose and carry out death sentences. The day after this plan became known, the conservative columnist William Safire wrote in the *New York Times* (November 15, 2001), "Misadvised by a frustrated and panic-stricken attorney general, a president of the United States has just assumed what amounts to dictatorial power to jail or execute aliens. Intimidated by terrorists and inflamed by a passion for rough justice, we are letting George W. Bush get away with the replacement of the American rule of law with military kangeroo courts." It later became known that the entire panoply of extraordinary powers the president gave himself by fiat were actually hatched in the office of Vice President Dick Cheney. As applied to aliens apprehended on American soil, the plan that concerned Safire abandoned a deeply ingrained principle of Anglo-Saxon law according to which whoever is on the territory of the king is at once subject to the king's laws and due the rights of the king's subjects. As applied to the Afghan battlefield, the plan arose in response to an obvious quandary: as al Qaeda fighters, and perhaps even bin Laden himself, were captured rather than killed, would they be criminals? war criminals? prisoners of war? The juridical procedures for dealing with them would present various complications and uncertainties in each case. The dilemma was simple. On the one hand, bin Laden would not have the protected status of a military or political leader in a war between nations, since he was not acting on behalf of any nation. Stated negatively, his death would not interfere with securing any truce, treaty, or postwar negotiation. On the other hand, to capture and then kill him or his followers would be outright murder. In opting for dictatorial powers and kangaroo courts, Bush sought to elude the dilemma at the expense of the rules of war as well as the rule of law. The actions he authorized at Guantánamo and elsewhere—as well as the unauthorized ones at Abu Ghraib—followed more or less directly from this posture.

Why would an American president want to enjoy this kind of arbitrary power? What daimon seized him?

First, the ideological interpretation. While there is no legal precedent in American history for the arbitrary power Bush gave himself as commander in chief, which is why Safire rightly and unrhetorically called it dictatorial, there are political antecedents. For just as law develops from precedent, political movements and governments often act in imitation of models—usually at the expense of their own legitimate aims. And though Safire was not wrong to call the plan a "Soviet-style abomination," there are models much closer to the ideological heart of Bush's own administration and party. In giving himself the authority to have suspected terrorists seized on American soil, removed to a ship on the high seas—or, as it turned out, to prison camps outside the United States, like Guantánamo—tried in secret, and executed, wasn't he creating a North American equivalent of the practice of Chile, Argentina, and Uruguay in the 1970s of "disappearing" presumed revolutionaries? Aren't the model interrogators the Latin American security forces and death squads who got their training at Fort Bening in Georgia? Aren't the model dictators those whom Nixon and Kissinger supported throughout the "dirty wars"? Shooting or hanging the culprits at sea would be a scarce refinement on pushing them from helicopters over the same seas.

Such a justice of ultimate ends is a nightmare that has threatened to eviscerate every principled reason for a war against terrorism, not least because it encouraged abuse and torture at Abu Ghraib and elsewhere.

So the question remains: why would an American president want to enjoy, how could an American president in fact enjoy, such power?

I will venture a moral interpretation of George W. Bush's daimon: Before Bush became president of the United States, his sole experience of the vocation of politics and life-and-death decision was as the governor of Texas. He oversaw the execution of 152 prisoners in six years.

Texas law is unlike most states'. While the governor has the duty of carrying out death sentences like all governors in states with the death penalty, his power to stop an execution, even through an act of clemency, is severely limited. There is scant evidence of Bush's actual experience of this predicament, except for the telling evidence that he never voiced any awareness that it was a predicament. A hundred fifty-two times he issued the order to execute someone while scarcely possessing the authority to stop it. Texas suspends the ethic of responsibility that governors and presidents are traditionally made to assume through their power to grant clemency or commute sentences; they are given that power because they are at once the highest authority in the executive apparatus of the state *and* answerable to the political and historical judgment of the people for whatever decisions they make. Gerald Ford decided to face the judgment of his fellow citizens when he pardoned Richard Nixon for the crimes he may have committed as president. Texas gave George W. Bush no such burden. As a governor of Texas, he was invested with the power over life and death without responsibility for life or death. The power he was charged with exceeded his responsibility; power alleviated him of responsibility.

George W. Bush's initial embrace of politics as vocation inured him to the full weight of the ethic of responsibility. It is not an inappropriately ad hominem argument to say so, because in never questioning, never protesting, never opposing, never seeking to reform the authority with which he was empowered by Texas law, Bush revealed a *public* moral flaw.

The enjoyment of power is inescapably ambivalent. When Weber identified one of the "inner enjoyments" of the political vocation as "the feeling of holding in one's hand a nerve fiber of historically important events," he brought out the nature of the ambivalence. To hold power is at the same time to be gripped *by* power. The politician does not simply enjoy power, he is enjoyed—seized, jolted, thrilled—*by* power.

Bush's experience as the governor of Texas throws more light on the flaws and risks of his particular hold on power than do many of the other, more obvious factors: he is the son of a former president, the weaker son according to family legend; he attained the White House without a mandate or majority support and thanks to the casuistry of a Supreme Court appointed by his father and his father's patron; he shored up his admitted deficiency in foreign affairs by choosing a vice president, secretary of state, and national security adviser who achieved their credentials in his father's and the patron's administrations. Those inauspicious beginnings of his presidency are not the decisive features of his political physiognomy. They simply define the context in which he—and he alone—made and had to make ultimate decisions.

The question that weighed on the perilous undertakings of his presidency—the war against Islamic terrorism, the Israeli-Palestinian conflict, Iraq—lies elsewhere. For the daimon presiding over this politician appeared when he habituated himself to a discrepancy between power and responsibility. While "whoever wants to engage in politics at all," as Weber said, "lets himself in for the diabolical forces lurking in all violence,"[1] Governor Bush let himself be exonerated from ultimate consequences when in the grip of ultimate power.

The New American Exceptionalism

The power-beyond-responsibility was not only a temptation for a new president of the United States; it had been the nation's actual situation since the end of the Cold War. The sole superpower had not asked itself for what ends and with what responsibilities it held such power. A decade had passed without the body politic troubling itself over such a historic question. George W. Bush took a bold step toward overcoming this inertia in planning war against Iraq. He advanced an answer to the historic question of the post–Cold War in *The National Security Strategy of the United States,* the document transmitted to Congress in September 2002. He called upon the nation to assume supreme global power. Such an aspiration is without precedent in American history. Even the most ambitious assertions of American power—the Monroe Doctrine, the Spanish-American War, the Cold War itself—did not aim for global supremacy. The new vision contained three declarations giving the United States a unique right over other nations: the right to "pre-emptive" war, the right to overthrow governments considered hostile, and the right to develop weapons outside any treaty structure and beyond any principle of deterrence.

---

[1]Max Weber, "Politics as Vocation," in *From Max Weber: Essays in Sociology,* ed. and trans. H. H. Gerth and C. Wright Mills (New York: Oxford University Press, 1946), pp. 125–26.

Each of these declarations is inimical to the ideals and efforts in international relations of the past half century. In taking this stance toward post–Cold War American power, the Bush administration broke with the consensus not only of the first Bush and Clinton administrations but also of the NATO partners and the European Union as a whole. Contrary to the situation faced after the September 11 attacks and in the war against al Qaeda and the Taliban in Afghanistan, the Bush doctrine transforms the legitimate right to self-defense into a potential license to aggression in the name of a self-declared self-defense. The claim to self-defense is always susceptible to abuse, as in the unfortunately forgotten example of the Gulf of Tonkin resolution that authorized President Johnson's escalation of the war in Vietnam. The new principle of preemptive war, or, more precisely, of unilaterally decided preemptive war, threatened to nullify all international constraints on the determination of legitimate self-defense. When Bush first tested the doctrine by declaring his plan to invade Iraq with or without UN approval, American allies around the world reacted in alarm that such a nullification would undermine not enhance international security. When Bush further tested the doctrine by invading Iraq without UN approval, the justification of this unilaterally decided preemption—Saddam Hussein's development of weapons of mass destruction—proved, almost immediately, to be utterly false. There were no weapons to preempt. The diplomatic costs of such bravado and error will not really become known for a long while, since the loss of other nations' trust is difficult to measure until some future international crisis that requires American leadership.

The extremist version of American exceptionalism embodied in the Bush doctrine admits of only one constraint on American power, namely, America's good intentions and values. That other nations, including most of our strongest allies, should find little reassurance in this is obvious enough and a strong reason to oppose the principles outlined in *The National Security Strategy of the United States*. But just as troubling is the consequence for American democracy of acting on these principles. In modern nations the people's interests are protected by the constraints imposed by international agreements, treaties, and alliances, since those constraints temper their own leaders' will and judgment. Modern democracies, modern masses in general, do not mobilize for war spontaneously. They are mobilized by their leaders, whose will and judgment are susceptible to all the pitfalls (and corruptions) of power. A mobilization for war also inevitably requires a strong dose of deception if not outright delusion regarding the motives and promised rewards of the battle. The human devastation of modern war has taught nearly every Western nation that international legal-diplomatic constraints protect it not only against the aggressive designs of other countries but also against its own.

Recognizing the warped principle and intrinsic danger of the Bush doctrine, let us also acknowledge that it was from the beginning a vision more than a realizable blueprint. The national security document swings between affirmations of the United Nations, multilateral alliances, and international law and bold claims to America's unique unilateral rights. The oscillations are not simply contradictory or hypocritical. They reflect the reality of an international situation that does not fit the hardliners' desideratum. The administration's engagement with the United Nations and its intricate negotiations with Russia and especially France over the Iraq resolution showed that a UN mandate and multilateral support were preferable to the unilateral aggression that the doctrine seeks to justify. The path toward a second Gulf War was a bit more crooked than labels like *hyperpuissance* and Empire suggest. So, too, as soon as the unauthorized invasion took place, the Bush administration was back at the United Nations to gain recognition of the occupation. And as soon as the occupation approached a crisis state, the administration renewed negotiations with the United Nations and NATO to come to its aid.

How, in the wake of a twelve-year absence of engaged political debate, did the United States find itself at the crossroad where its leader was calling for global supremacy? The red thread lies in the vicissitudes of the Powell doctrine in the post–Cold War. Even though the supposed hallmark of Donald Rumsfeld's Department of Defense was to overturn the Powell doctrine, what was really at issue was an overhaul of the means for accomplishing the same ends. The doctrine originated as a strategy for the effective use of conventional forces in major but relatively localized or regional conflicts. Defense policy set the goal of maintaining a military of professionalized troops, high-tech weapons, and air power at a level capable of decisively winning at least two major conflicts simultaneously with minimal American casualties. This post-Vietnam vision of American might was more ambitious than any undertaking since the development of nuclear weapons and intercontinental ballistic missiles, but even so it did not contemplate making America the dominant and sole global power. On the contrary, the streamlining and buildup of American forces that Colin Powell directed during the Reagan-Bush years presupposed that American power would remain counterbalanced by nuclear parity with the Soviet Union.

Since the first real test of the post-Vietnam military did not come until the post–Cold War crisis provoked by Iraq's invasion of Kuwait in 1990, the Powell doctrine proved its worth in just the sort of war but not the sort of world it was designed for. The swift victory in the Gulf War and the newness of the international-multilateral effort that the first President Bush achieved under the auspices of the United Nations and through an alliance of Western, Arab, and Eastern European nations seemed to cast the United States in the double role of sole superpower and global leader. The New World Order proved of course considerably more ambiguous, and the astonishing military success against Iraq clarified very little. President Bush won the war and lost re-election; Saddam Hussein lost the war and remained in power, defying his truce agreements with the United Nations despite the toll that this defiance exacted on the Iraqi population. Americans saw the extent of their military supremacy clearly displayed in 1991 but recoiled from the political task of comprehending their new responsibility as global leader. Clinton won the 1992 election in part by assiduously avoiding discussion of foreign policy. The Clinton years saw the emergence of another post–Cold War reality, namely, the civil wars and genocides carried out in the name of ethnic, religious, and tribal identities. Confusion and indecisiveness pervaded the American understanding of its global responsibility. Being the sole superpower shed no light on *how* to be the sole superpower.

In the absence of a new definition of American responsibilities, the understanding of what it meant to have military supremacy split during the Clinton interregnum along the faultlines of the Powell doctrine itself. On the one hand, the Clinton administration hewed to the doctrine's reluctance to put American soldiers into battle unless clear objectives could be achieved with minimal casualties; when eighteen American soldiers were killed in Somalia in Operation Restore Hope, which Clinton had inherited from the Bush administration, the new president reacted to the public's horror at seeing an American soldier's body dragged through the streets of Mogadishu by disengaging from the conflict. On the other hand, foreign policy conservatives spent eight years elaborating various ideas for America to use its military supremacy more actively and aggressively. The Republicans were bitter about being out of power for the first time in twelve years, but now that they were relieved of the responsibilities of power they were freer to nurture their dreams of American supremacy. The hardliners came to disdain the elder Bush's reliance on international law and coalitions, blaming the built-in constraints of multilateralism for the decision in 1991 not to pursue the war on Iraq all the way to the overthrow of Hussein; they repudiated involvement in crises like Bosnia or Rwanda as too distant from American "national interests"; and they intensified their Reaganite contempt for the United Nations, with the Republican majority in the Senate undermining various UN programs. Many out-of-power Republicans, including the future vice president and secretary of defense, deepened their ties to military-related industries and various foreign governments through their lucrative positions as CEOs, consultants, and lobbyists. The entrepreneurs and think-tank intellectuals developed the ideas that eventually created the national security strategy unveiled a year after September 11.

The Powell doctrine, forged from Colin Powell's own conviction that the war in Vietnam failed for lack of defined goals and public support, underlay both Clinton's timidity and the conservatives' hubris. The United States should use its extraordinary military might *only* in the pursuit of clear national interests; the United States has extraordinary military might *to use* for clear national interests. As the president and his antagonists followed the same principle along these opposite tracks, the body politic failed to ask, What *are* the United States's national interests at the dawn of the twenty-first century?

The Powell doctrine had in effect institutionalized the Vietnam Syndrome, inculcating in the government and the people this paradoxical timidity and hubris. The wealthiest and mightiest nation on Earth holds as its ideal of military force the capacity to devastate whole armies and bend other nations to its will with the barest risk of American lives. That military planning should aim to inflict the maximum effective harm on an enemy at a minimum of harm to oneself is merely rational. However, that a nation should be willing to commit itself only to those battles in which it will not suffer losses conditions it to timidity. That the same nation should also maintain overwhelming military superiority for the purpose of influencing world affairs requires considerable hubris. Timidity and hubris would simply be moral attributes by which to judge a nation's character more or less severely, but the combination of timidity and hubris, the relentless oscillation between them, jeopardizes American democracy and international relations.

For all the novelty of the Powell doctrine, the United States remained caught in the same self-created dilemma that Octavio Paz pinpointed on the eve of Ronald Reagan's election in 1980: "If they could, Americans would lock themselves up inside their country and turn their backs on the whole world, except to trade with it and visit it. . . . With respect to international affairs . . . , the positions of liberals and conservatives are interchangeable: both shift quickly from the most passive isolationism to the most determined interventionism, though these shifts do not substantially modify their vision of the outside world. It is not strange, therefore, that despite their differences both liberals and conservatives have been by turn interventionists and isolationists."[2]

How the oscillation affects American democracy was evident in the tenor of political debate and public opinion regarding a possible invasion of Iraq. Bush mobilized public support with the deceptive claim that the overthrow of Saddam Hussein was an extension of the war against international terrorism. The Democrats in the House and Senate gave their consent to war not because they shared Bush's vision of national security and American interests—clearly they did not—but because they feared appearing unpatriotic to the voters in November 2002 and hoped to turn the midterm elections into a referendum on Bush's dismal economic policies. The Democrats first tried to postpone the Iraq debate until after the election, as though the issues of war and national security should not be debated in front of the voters, and then reversed themselves and pushed the debate and congressional vote to a rapid conclusion in order to set aside the Iraq issue before the elections.

[2]Octavio Paz, "Imperial Democracy" [1980], in *One Earth, Four or Five Worlds: Reflections on Contemporary History,* trans. Helen R. Lane (New York: Harcourt Brace Jovanovich, 1985), pp. 36, 47.

"As Congress prepared to sign off on the war resolution," according to one report, Democrat Tom Daschle, the Senate majority leader, "sounded relieved, predicting that Americans would start brooding over the economy 'once we get this question of Iraq behind us.' "[3] The Democrats of course actually put Iraq in front of us and in front of the American soldiers and vast number of Iraqis destined to die in the conflict. Their self-serving calculus deprived the polity, yet again, of a decisive debate on America's global role.

Meanwhile, even as the Republicans were mobilizing the masses and the Democrats were hedging their bets, the electorate itself demonstrated the ambivalence toward war and peace that haunted the political elites of both parties: pollsters disclosed that while a vast majority of Americans favored overthrowing Saddam Hussein, their support for the war dropped precipitously if the anticipated American causalties rose to a certain level. In one of the more obscene manifestation of the Vietnam Syndrome, pollsters were able to neatly quantify the unacceptable level of American losses in a war with Iraq: five thousand. The public's otherwise enthusiastic support for invading another country, probably devastating armies and cities and killing thousands of civilians, dropped to a mere 33 percent if it were to cost five thousand American lives. Hubris and timidity, timidity and hubris. The separation of power from responsibility afflicted the entire body politic—the Republican administration, the Democratic opposition, and the electorate.

The Cold Warrior Myth

The hardliners' hubris was rooted in the Reagan-Bush years when Dick Cheney, Donald Rumsfeld, and Condoleezza Rice formed their vision of American foreign policy. Their experience of the final decade of the Cold War and the beginnings of the post–Cold War left them with a myth. They believe that Reagan's foreign policy caused the collapse of Soviet communism. The several causes—decades of dissidence, Gorbachev's glasnost and perestroika reforms, the political and economic aspirations of Eastern Europeans and the reawakening of their long-suppressed liberal traditions, the Soviet debacle in Afghanistan, the legitimation crisis

[3]Frank Rich, "It's the War, Stupid," *New York Times*, October 12, 2002.

of the Soviet state as the people withdrew even their passive allegiance—
are in the eyes of these ex–Cold Warriors mere epiphenomena compared to Reagan's assaults on the Evil Empire: the invasion of Grenada,
the complicity with death squads in El Salvador, the illegal funding of
the Contras in Nicaragua, and a nuclear strategy, including the development of tactical weapons and the Star Wars research, that undermined the deterrence doctrine (Mutually Assured Destruction) in favor
of nuclear superiority. The Reaganites consider *themselves* the victors of
the Cold War.

Their self-proclaimed victory left them, however, in need of a new
myth. What adversary was there to replace the Soviet Union? The theme
of the missing adversary has been sounded with various meanings.
"When You've Lost Your Best Enemy" is the title of the chapter in Powell's autobiography where he deals with the months in 1991 that saw
Iraq's invasion of Kuwait follow quickly upon the historic negotiations
between the United States and the Soviet Union that effectively ended
the Cold War. Whereas Powell saw irony in the coincidence and signs
that the post–Cold War would create crises of a new type in the absence
of the U.S.-Soviet balance of power, many conservatives began searching
the horizon for an enemy to replace Soviet communism. The think-tanks
urged a harsher policy toward "rogue states," as Clinton called Iraq,
North Korea, and Libya, and also argued against further rapprochement
with China. Samuel J. Huntington's thesis of the *clash of civilizations* persuaded a large circle of foreign policy conservatives well before September 11 that the West was entering an era of potentially violent conflict
with Islam. Some conservative commentators openly admit that Islam
answers the need for a substitute for communism. Stanley Kurtz, writing
in the journal of the Hoover Institute, summarizes Huntington un-
apologetically in such terms: "Humans require identity, and they acquire
it, says Huntington, through the enemies they choose. With the collapse
of Cold War enmities, new forms of identity will inevitably be con-
structed upon new patterns of hostility. Differences of religion and culture will provide the needed template for the clashes to come."[4]

Whether criticized as an invitation to racism or defended as some
kind of psycho-metaphysical truth, the thesis that America is at war

[4]Stanley Kurtz, "The Future of 'History,'" *Policy Review* 113 (June 2002).

with Islam is ill-founded. September 11 revealed the scope of Islamists' hatred of the West, but the American response was not the beginning of a war on Islam. Bush's *National Security Strategy of the United States* explicitly repudiated the idea of a clash of civilizations. The repudiation was significant because it took a position against such an influential strand of conservative thought. Many critics saw in it simply a whitewashing of the administration's true beliefs and motivations. I see it, rather, as confirmation of the fact that the Bush doctrine and the policy toward Iraq originated in thoroughly secular motives and ideas.

The hardliners adhered to a threefold program from the beginning of the Bush administration: permanent military buildup, isolationist-unilateralist diplomacy, and a Kissinger-style understanding of national interests. These are ideological habits of the Cold War, reinforced by the Bush officials' business experience in defense-related corporations. What is crucial, though, is that these three commitments do not cohere as a policy *unless* the United States faces some intractable adversary. The conservatives have an obvious nostalgia for the bipolar world of the Cold War, and more particularly for the glory they have granted themselves as the victors. Nevertheless, it was their adherence to this threefold program that sent the new Bush administration in search of a new adversary even before September 11. The provocative probing of Chinese air defenses resulted in the capture of an American plane, and the decision to end the American role in establishing a dialogue between North and South Korea seemed to mark a preference for maintaining rather than ameliorating North Korea's status as a dangerous nuclear parriah. The sheer irrationality of this attitude suggested that the underlying rationale may well have been simply to justify development of the so-called missile shield.

Secretary of Defense Rumsfeld pushed the hard line of isolationist-unilateralist "national interest," including the repudiation of involvement in Bosnia, the abrogation of the anti-ballistic missile treaty, the pursuit of the new Star Wars initiative with or without allies' consent, and the supremacy of American prerogatives within NATO and over the European Union.

The other hardliner, Vice President Cheney, is especially steeped in the Reaganite politics of ultimate ends. If one looks behind the CEO who parlayed his defense contacts into a personal fortune and behind

the secretary of defense who orchestrated George Herbert Walker Bush's
Gulf War, one sees the Wyoming congressman of the 1980s who disdained open debate and democratic scrutiny of the means and ends of
state-sponsored violence. "When the Iran-Contra scandal broke," Congressman Cheney "became the leading defender of the Reagan Administration's funding of the Contras in violation of a congressional vote.
He served on the special Iran-Contra investigating committee and
supervised the vigorous defense of Presidential authority that the committee's Republicans issued. . . . Cheney thought Reagan, as a matter of
right, should have been able to support the Contras without having to
clear it with Congress."[5] He thus effortlessly became the leading propo-
nent of secret tribunals after September 11.[6]

Aggressive new weapons programs, unilateralism, isolationism as regards "nation building" (when it comes to genocidal "tribal" conflicts)
combined with interventionism (when it comes to "national interests"),
unlimited presidential power in fostering "regime change" abroad—
these were the ideological building blocks of the Bush administration
from the outset, a thoroughly secular and materialistic ideology.

Kant with Arendt

The intellectual vision behind the Bush doctrine finds expression in the
Great Powers thinking of Robert Kagan, whose contrast of Americans
and Europeans as Mars versus Venus or Hobbes versus Kant became a
staple of discussions of the transatlantic diplomatic conflicts over Iraq.
Kant is continually caricatured in the Hobbes-versus-Kant motif. He
did not hold a paradisical view of the relations among nations. On the
contrary, his reflections in "Perpetual Peace" assume that nations are in
short mean and brute, no matter how liberal and lawful their internal
governance: "Although it is largely concealed by the governmental constraints in law-governed civil society, the depravity of human nature is
displayed without disguise in the unrestricted relations which obtain
between the various nations." His view of the relation between nations
was in this sense *darker* than Hobbes's, for the author of *Leviathan*

[5]Nicholas Lemann, "The Quiet Man," *New Yorker,* May 7, 2001, p. 66.
[6]*New York Times,* October 24, 2004.

considered the sovereign to be primarily devoted to the "Peace and Security" of the commonwealth, whereas Kant had already seen that nations gone to war were capable of unconstrained destructiveness. Kant valued international laws and treaties because no nation's civil order, no matter how *exceptional*, intrinsically tempers its capacity for violence and brutality toward other nations. A peaceful international order would have to be "created and guaranteed by an equilibrium of forces and a most vigorous rivalry" among nations. Their rivalry would come from "linguistic and religious differences," a rivalry that could be not only tamed but also enjoyed "as culture grows and men gradually move toward greater agreement over their principles." Civilization *is* the clash of civilizations, according to Kant. Acutely aware that his ideas were ahead of his time—as they remain painfully ahead of our own—his commitment to international agreements is practical, not utopian: "And while the likelihood of its [perpetual peace's] being attained is not sufficient to enable us to *prophesy* the future theoretically, it is enough for practical purposes. It makes it our duty to work our way towards this goal, which is more than an empty chimera."[7] What is an empty chimera is the neoconservatives' tautology *power is power* and the resulting illusion that American might unambiguously affords American democracy protection, and it is with this tautology and illusion that the Bush administration turned the United States away from the goal of a lawful equilibrium among nations.

Hannah Arendt challenges the idea of power that is contained in the supposedly self-evident phrase "great powers." Military might does not necessarily protect the democracy that possesses it. Arendt's notion that power "corresponds to the human ability not just to act but to act in concert," while might is merely the capacity for violence and violence itself is but the *breakdown* of power, was meant to underscore that power is the foundation of political community and as such is an end in itself. She then offers the following clarification in order to arrive at the question of what justifies the use of force:

> This, of course, is not to deny that governments pursue policies and employ their power to achieve prescribed goals. But the power struc-

---

[7] *Kant's Political Writings*, ed. Hans Reiss and trans. H. B. Nisbet (Cambridge: Cambridge University Press, 1970), pp. 103, 113–14.

ture itself precedes and outlasts all aims, so that power, far from being the means to an end, is actually the very condition enabling a group of people to think and act in terms of the means-end category. . . .

Power needs no justification, being inherent in the very existence of political communities; what it does need is legitimacy. . . . Power springs up whenever people get together and act in concert, but it derives its legitimacy from the initial getting together rather than from any action that may follow. Legitimacy, when challenged, bases itself on an appeal to the past, while justification relates to an end in the future. Violence can be justifiable, but it will never be legitimate. Its justification loses in plausibility the farther the intended end recedes into the future. No one questions the use of violence in self-defense, because the danger is not only clear but also present, and the end justifying the means is immediate.[8]

To summarize Arendt: Violence in self-defense loses its justification the less immediate and clear the danger. Violence itself marks a breakdown, not an extension, of the power embodied in political community. Force, no matter how superior, never guarantees an increase in power; it can be, and historically often has been (her example is the American defeat in Vietnam), defeated by an enemy with less strength but greater solidarity. To summarize Kant: no nation's civilized attainments or enlightened laws prevent it from barbarity when its relations with other nations are unconstrained by international agreements.

Such are the historical and political reflections that ought to guide a National Security Strategy of the United States. It is even conceivable that they might have become ingrained into the American understanding of its extraordinary military might after the Cold War, for they are not after all incompatible with what is known of Bill Clinton's attitudes toward democracy and arms, political community and military force, national interest and foreign affairs. Clinton failed to effectively articulate such a new understanding, and this intellectual and political vacuum was quickly filled by the Republicans. Their vision of American foreign policy turns the Kantian-Arendtian perspective upside down. It equates power and might, believes that the virtue of American democracy

[8]Arendt, "On Violence," pp. 143, 150–51.

guarantees the rightness of American force, justifies violence as self-defense in the absence of immediate and clear danger, and exempts America from the laws it would impose on every other nation.

The Bush doctrine exacerbates a danger inherent in the Powell doctrine. The Powell doctrine was designed to avoid military engagement except where the results could be decisive and swift, and the armed forces expanded to be able to meet that standard in many situations. Arms can seem simply instruments of policy, but they also shape it. From the diplomatic standpoint, the Powell doctrine makes the government prone to avoid conflicts that otherwise might be considered urgent and justified. Hence Clinton's hesitation in Somalia, Rwanda, and Bosnia, and the Republicans' own disdain for "nation building." But the Powell doctrine also tempts American leaders to *seek,* even foster, conflicts where the preferred sort of military engagement is possible. A preferred strategy needs fit enemies.

Therein lies the most plausible explanation for the fact that Bush took the first great occasion after the fall of the Taliban, his State of the Union Address in January 2002, to shift the entire focus of his foreign policy onto the "axis of evil." Possible conflicts with North Korea and Iran and especially Iraq seemed better suited to the Powell doctrine than the war against terrorism. The overthrow of the Taliban had been accomplished within the doctrine's strategic framework of massive air power and minimal commitment of ground troops only because the militias of the Northern Alliance could be relied on to do the heavy fighting. Beyond that first phase, however, the struggle against al Qaeda and international terrorism required a complex multilateral effort of police work, a long-term engagement in Afghanistan to stabilize the country and break up al Qaeda and the Taliban, and a still largely neglected program to foster economic and political reform in several Arab and other Muslim countries. Neither the Bush doctrine nor the Powell doctrine was suited to such tasks. Even as the administration revealed that its primary aim was to prepare for "regime change" in Iraq, it plainly ignored the mounting evidence that the war in Afghanistan had not neutralized al Qaeda: in June 2002, anonymous "senior administration officials" told the *New York Times* that "classified investigations of the Qaeda threat now under way at the F.B.I. and C.I.A. have concluded that the war in Afghanistan failed to diminish the threat to the United

States. . . . Instead, the war might have complicated counterterrorism efforts by dispersing potential attackers across a wider geographic area" (June 16, 2002); in September, just as the Bush administration was taking its Iraq resolutions to Congress and the United Nations, a UN study reported that bin Laden's network was substantially intact and active; and the bombing of a night club in Bali in October confirmed that groups loosely affiliated with al Qaeda could carry out major attacks of their own. Iraq must have seemed a tangible enemy, in contrast to the hydra of Islamic terrorism. It was fixed to its territory and ruled by a tyrant unlikely to survive a military conflict with the United States. This was the sort of war our military was built to win. So was born Shock and Awe.

Targeting Iraq

I am not suggesting that the Bush administration fabricated Iraq, North Korea, and Iran as adversaries. All three countries were thought to be developing weapons of mass destruction, and all three regimes were capable of reckless and dangerous acts. The United States had ample reason to lead a concerted international effort to undo their development of nuclear, biological, and chemical weapons and missile systems. And there should be no doubt that such international efforts usually depend on American leadership. At issue, rather, are the means and justifications that Bush chose, and the specific perils they posed. Bush justified the confrontation with the axis of evil as an extension of the war against terrorism despite the lack of any evidence connecting al Qaeda and North Korea, Iran, or Iraq. There was ample evidence that terrorist networks seek to acquire nuclear, biological, and chemical materials, but these countries were unlikely suppliers. As became dramatically clear in the months that followed, Pakistan had been all along the most dangerous source of proliferation. Moreover, contrary to the World War II-vintage image of an *axis* of evil, the three countries presented the United States and the world with quite different problems.

The international projects having the most bearing on the conflict that exploded onto the world stage on September 11—the struggle against al Qaeda, rebuilding Afghanistan, and an Israeli-Palestinian peace—*receded* in importance for the Bush administration in favor of a confrontation with the axis of evil. It had no plan regarding Iran, and its approach to North Korea was geared to sustaining rather than resolving the crisis. The aggressive unilateral posture came down to its long-held desire to invade Iraq and overthrow Saddam Hussein.

The meaning of this intention had undergone a change since September 11. Beyond the earlier desire to "finish the job" left undone by the first President Bush in the Gulf War, the war against Iraq became a part of the larger aim of demonstrating, displaying, instantiating, America's unilateralist right to wage preemptive war and overthrow hostile governments. It became a means to the end of establishing the new exceptionalist principle. One of the greatest, if slow-developing, dangers of the Iraq intervention to American democracy itself lies in this reversal of ends and means. Was it a war ultimately undertaken to justify the principle of preemptive war and demonstrate America's global supremacy? The question is impossible to answer directly. However, since the stated reasons for the war—Saddam Hussein's weapons of mass destruction and ties to al Qaeda—proved half-illusory, half-invented, it is difficult to distinguish the actual from the fraudulent motives for the intervention.

The man for whom the Powell doctrine is named is widely believed to have represented the countertendency to neoconservative and hardline views in the Bush administration. His was indeed the most audible voice stressing multilateralism over unilateralism, arduous diplomacy over precipitous aggression, international law over exceptionalist right. However, there is also a pattern, and latent limit, to Powell's perspective. As I have argued, the hardliners' view is ultimately based on their threefold commitment to permanent arms development, unilateralism, and a narrow conception of national interests. Powell shares the assumptions on military supremacy (his entire career has been devoted to developing it) and national interest (his own reservations about Somalia, Bosnia, and Rwanda did not differ from Rumsfeld's[9]). So long as Powell's com-

---

[9]Writing in 1995: "we are currently witnessing the chaos that occurs when states revert to anarchy, tribalism, and feudalism, as in Somalia, Rwanda, Burundi, Liberia, and Sierra Leone. Television delivers tragic scenes from these places into our living rooms nightly, and we naturally want to do something to relieve the suffering we witness. Often, our desire to help collides with the cold calculus of national interest. In none of these recent foreign crises have we had a vital interest such as we had after Iraq's invasion of Kuwait and the resulting

mitment to international negotiations and multilateral coalitions did not lead him to challenge these other two assumptions, he seldom prevailed against the hardliners' determined unilateralism, and when he failed to persuade he remained loyal to the president who rejected his advice.

Al Qaeda and Ultimate Ends

Did Islamic terrorism and the axis of evil step into the role of supreme adversary after September 11? This question is not as easy to answer as it might appear, given Bush's recurrent recourse to the rhetoric of moral absolutes and ultimate ends: first a *crusade* against terrorism and *Infinite Justice,* the battlecries that were quickly abandoned for the sake of Arab and Islamic sensitivities, and then the *axis of evil* and the mission to "rid the world of evil." Opponents of the war against terrorism seized on Bush's slogans to claim that the United States had, in Lewis Lapham's words, embarked on an American jihad. Edward Said, writing in *Le Monde* (June 25, 2002), claimed that "The American administration is controlled by the alliance of Christian fundamentalists and the pro-Israeli lobby." Toni Negri dubbed the war in Afghanistan a conflict between "the Taliban of fundamentalism" and "the Taliban of capitalism."

Neither Bush's slogans to justify American policy nor the critics' slogans to denounce it capture the motives and substance of the policy itself.

Negri mistakes his own wit for analytical insight. There is no political parallel or moral equivalence between the Bush administration and the Taliban, because there is no parallel between a democratically elected government and a theocratic junta and no equivalence between capitalism and Islamic fundamentalism. Capitalism is the historical result of the more or less coordinated, more or less uncoordinated actions of

---

threat to Saudi Arabia and the free flow of oil. These later crises do not affect any of our treaty obligations or our survival as a nation. Our humanitarian instincts have been touched, which is something quite different. Americans are willing to commit their diplomatic, political, and economic resources to help others. . . . But when the fighting starts, as it did in Somalia, and American lives are at risk, our people rightly demand to know what vital interest that sacrifice serves." Colin Powell, with Joseph E. Persico, *My American Journey* (New York: Ballantine Books, 1996 [1995]), p. 589.

countless nations, individuals, enterprises, and movements, including workers' movements, over the last four centuries, while Islamic fundamentalism is the programmatic yoking of religious doctrine, anti-modernity, male supremacy, and violence. Negri is wrong for the same reason Huntington's devotees are wrong: September 11 did not represent Islam versus the West, but a fanatical strand of Islamic fundamentalism against modernity, democracy, and secularism.

Said, despite his own lucid criticism of *The Clash of Civilizations*,[10] succumbed to Manichaeanism himself when he postulated that some diabolic coalition of Christian fundamentalists and rightwing Zionists controls the American government. The facts are alarming enough. Christian fundamentalists are a crucial element of the electoral bloc of the Republican Party, and an influential network of defense officials and intellectuals came into office sharing Ariel Sharon's then apparent resolve never to permit a Palestinian state. But it was an error to conclude that they *are* the government in power, and an even greater error to suppose that the linchpin of American support for Israel is religious conservatism. The American commitment to Israel is deep and cuts across the political spectrum. Sharon was accepted first and foremost because he had been elected Israel's leader, and he was viewed with caution and suspicion in the United States as by the Israelis who elected him in the wake of the failed peace talks between Barak and Arafat and the suicide bombings that were the hallmark of the second intifada. Israel's ordeal with suicide bombers caused Americans and Israelis to give Sharon wide latitude in countering the attacks, too wide a latitude in my view, but by the same token his original policies toward the settlements and occupation of the West Bank were continually criticized by the American media and often by the Bush administration itself. American policy is anchored in decades-old commitments to the survival of Israel, the demand that its existence be recognized by the other Middle Eastern nations and the Palestinians, and support of its struggle against terrorism.

The American response to September 11 cannot be understood without fully acknowledging the uniqueness and threat of the sort of terrorism practiced by al Qaeda. The terrorists are truly devoted to an

[10]Edward Said, "The Clash of Ignorance," *The Nation*, October 21, 2001.

ethic of absolute ends. Suicide bombers are indoctrinated to believe that their own deaths are a sacred martyrdom and that the slaughter of innocents is divine justice against infidels. Sunni Islamism transcends the contradiction that Weber found in Western variants of the ethic of ultimate ends. His archetype of the absolute ethic is the Sermon on the Mount, and the efforts to impose it on society that most concerned him were those of revolutionaries devoted to ideals of universal brotherhood and perpetual peace. The adherents of the earthly realization of moral absolutes expose themselves to a labyrinth of consequences in turning to violence as a means to ultimate ends. However, since politics cannot rule out violence as a means, its ethic of responsibility requires at once a sense of proportion and an awareness of tragedy; when, conversely, the ethic of absolute ends seeks to impose itself on human society, it not only dislodges the ethic of responsibility but also drowns its own ideals in violence, deception, and self-deception. "He who seeks the salvation of the soul, of his own and of others, should not seek it along the avenue of politics, for the quite different tasks of politics can only be solved by violence. The genius or demon of politics lives in an inner tension with the god of love. . . . Everything that is striven for through political action operating with violent means and following an ethic of responsibility endangers the 'salvation of the soul.' "[11]

Islamist terrorism escapes this inner tension between violence and absolute good. The death of infidels and the paradise of martyrs *are* sacred ideals. When Americans came face-to-face with this avatar of the ethic of ultimate ends, our fear and trembling resonated perhaps with the uncanny echo of Christianity's not so distant glorification of crusaders and martyrs, devout killers and sanctified victims. *Crusade! Infinite Justice! Absolute evil!* These absurdities could roll off the president's tongue only because they came from somewhere a bit deeper than his own psyche. Even though the rhetoric of moral absolutes persisted, including the absurd claim in the National Security document that the purpose of American foreign policy is to "rid the world of evil," American policy under Bush did not launch a religious war, a clash of civilizations, or a capitalist jihad.

[11]Weber, "Politics as Vocation," p. 126.

A Grammar of Motives

How then to interpret Bush's constant recourse to moral hyperbole and messianic melodrama to justify the secular-materialistic purposes and principles of his administration's foreign policy? There is not a simple answer to this question. At least three not easily synthesized lines of interpretation are all plausible.

The first interpretation hews to a strict distinction between *motives* and *justifications* when it comes to analyzing political decisions. Foreign policy is driven by complex motives—from economic and geopolitical national interests to partisan political considerations and special-interest agendas to long-range goals that are fraught with uncertainty and are frequently simply ill-defined. Politicians in power often, perhaps usually, find it inexpedient to spell out all the motives as they try to justify a particular course of action. As Bush prepared the war against terrorism and the intervention in Iraq, he had to take stock of the electorate's limited willingness to support large-scale, sustained military engagements. American political elites have faced this problem since the Vietnam War. In fact, Americans have always been reluctant to mobilize for military action. They are by instinct isolationist despite the long history of American interventions. When they do support military action, they typically have little understanding of the international scene, and they do not follow their leaders into war unless the justification is in the name of freedom.

Every leader is thus tempted to attach the claim of defending or extending freedom to his policies to legitimize them. Defending or spreading freedom is the messianic trigger of public support for military engagement. However just the cause of freedom may be, it is as easily manipulated as any other reason, just or unjust, for going to war. September 11, like Pearl Harbor, was a direct attack, and Americans supported the president's decision to overthrow the Taliban and pursue al Qaeda in Afghanistan as a necessary and justified act of self-defense designed to break up the terrorist organization that had perpetrated the attack. The Untied Nations and NATO held the same view. Neither the military offensive nor the overwhelming public support stemmed from demonology. Neither the action nor the support *needed* demonology.

Bush nevertheless retained the good-versus-evil rhetoric, perhaps already anticipating the more arduous task of persuading the public to an invasion of Iraq. For, no matter how honest the administration's error in overestimating Iraq's weapons of mass destruction because of faulty intelligence, it did know three things about Iraq: Iraq was not a direct threat to the United States, it had not been involved in September 11, and it had no viable ties to al Qaeda and scant interests in common with radical Islamism. Self-defense and the war on terrorism were thus the *justifications* for the war in Iraq but not the *motives*. The motives had more to do with the grandiose vision rampant among the neoconservatives at the Pentagon and the White House that getting a democratically elected, pro-American leader of Iraq would cause Muslim countries throughout the Middle East to fall like dominoes into the democratic camp. Americans were not ready to embrace this vision as their own. As Bush shifted the focus to Iraq, he muted the motive of remaking the Middle East and trumpeted a confrontation with the axis of evil, first and foremost Saddam Hussein's Iraq.

The rhetoric of absolute Good and Evil plugged the hole in a political discourse that had failed between 1989 and 2001 to persuasively define America's responsibility in relation to its power. One real danger of Bush's recourse to melodramatic and messianic rhetoric lay in the fact that the effectiveness of such rhetoric is actually very thin. The Bush administration was still in the grip of the Vietnam Syndrome: they remembered that the public revolted against a costly military engagement in Southeast Asia, but they forgot that the only justification that the Kennedy and Johnson administrations put forth for that war was another melodrama of good and evil: the evil of monolithic communist expansion. The brutal realities of the war in Vietnam eroded not just the polity's willingness to make sacrifices but also, and more decisively, its belief in the justification.

The second interpretation of the messianic posture concerns the role of the religious Right as an indispensable element of the Republican Party. Bush began his own association with evangelical activists when he reached out to them on behalf of his father's electoral campaigns of 1988 and 1992. During his own presidency, he kept them in tow with several significant but partial concessions to their agenda. His willingness to link his religious faith and his foreign policy was at first but another symbolic gesture of solidarity with his electoral "base." However, this exploitation of religious faith for political purposes had a much wider effect as it exacerbated and increasingly legitimized what evangelical activism had been tempted to do for years, namely, openly breach the boundary between private and public when it comes to religion. Bush contributed directly to this tendency with his faith-based initiatives that funneled substantial federal funds to evangelical groups, blurring the distinction between civic and religous commitments.

Evangelicals no longer hid their religious organizations' political objectives, and they no longer held back from religious expressions of their politics, as they had in the 1990s when, for example, they supported "stealth candidates" in local school board elections. The prevailing American attitude has long been one in which your religious convictions are expected to sustain, even determine your political opinions, but the *arguments* you make for your political stance are thoroughly secular. That is, you address arguments to your fellow citizens without reference to your religious convictions or theirs. It is part of civic respect for the religious convictions of others that public matters are not debated on explicitly religious grounds.

It is this reserve and respect that was aggressively cast aside in the 2004 campaign. Voters would say that since they were Christian they were voting for Bush. Such ignorance of the meaning of "Christian" is widespread among evangelicals, but more striking is the willingness to link religious and political affiliation. Religious figures from the archbishop of Denver to rural preachers in Ohio hammered away at abortion rights, gay marriage, and stem cell research. Churches became polling places in unprecedented numbers, partly based according to local officials' claims of security concerns about using schools as is the predominant tradition.

The 2004 presidential campaign saw the boundary between religious conviction and political belief, between a private domain of conscience and a public realm of citizenship, repeatedly breached. Conservative Catholic bishops made their thinly veiled endorsement of Bush over Kerry by claiming that any Catholic who *voted* for a candidate who supported abortion rights, gay marriage, or stem cell research was committing a *sin* and should not receive communion without confessing their sin. Voting Democratic or voting for Arlen Specter became a sin. The confusion of confessional booth and voting booth violates an essential democratic tenet, for the very existence of the secret ballot declares that no authority, religious or secular, is licensed to know how one votes. Meanwhile, evangelical leaders aggressively broke down all distinction between religious authority and political judgment on the basis of scandalously ignorant theology. According to a news report on the eve of the election, "the evangelical group Focus on the Family released 'a must-read election message' signed by its influential founder, James C. Dobson, and more than 80 prominent evangelical Protestants arguing that the Bible teaches lessons about proper government, including not only opposition to abortion and same-sex marriage but also support for preemptive military action against suspected terrorists and looser environmental regulations."[12] Sermons on Jerry Falwell's website cited the commandment against coveting thy neighbor as an injunction against taxes. Evangelical preachers have become committed politicians and electoral activists, and the Bush campaign did not hesitate to solicit evangelical churches to prosletyze on Bush's behalf.

As I look at the evolution of Bush's messianic rhetoric in relation to electoral politics, it seems to me that his religiously inflected justifications for the use of armed force, especially in Iraq, were at first a kind of signal of shared belief with an important segment of his constituency. The effect of this sort of gesture, backed by the concerted efforts of his advisers and campaign managers to use churches as an organizing tool, more and more emboldened religious conservatives to breach the barriers between religion and politics. In turn, Bush ratcheted up the messianic rhetoric of his foreign policy. Another trajectory ultimately fused with this spiraling escalation of religious fervor. For it was after—and only after—all the world could see that Iraq had had no weapons of mass destruction and no ties to al Qaeda and September 11 that Bush turned to making the end of tyranny and spread of democracy the primary justification for the war and occupation. As he did so, and with ever greater intensity as the occupation itself confronted the evils of the insurgency, he gave the freedom-versus-tyranny theme an ever more messianic flavor.

The third interpretation of Bush's messianic justifications of war and

[12]David D. Kirkpatrick, "Battle Cry of the Faithful Pits Believers Against the Rest," *New York Times,* October 31, 2004.

occupation (and even torture) comes back to the question of his experience of power; ad hominem though the argument may be, it still refers to the public man, not the private. By temperament and circumstance, George W. Bush as president has let himself be driven to decisions by the mere existence of the military might at his disposal. In Washington as in Texas, his grip on power put him in the thrall of power, a power that by its very nature overflows responsibility. I do not doubt—indeed I fear—that his own religious convictions have in fact stirred within him a deeply felt sense of messianic purpose that he grafts onto his vision of America's role in the world. If so, it fits the pattern of his embrace of the political vocation: if a politician is seized by power that exceeds his own capacity for judgment, overwhelming his sense of ultimate responsibility and eclipsing any awareness of tragedy, where does he turn when he must decide life and death?

... we promise all due submission and obedience.
—MAYFLOWER COMPACT

THE IMAGINATION OF POWER

State of Exception

Fear and hubris have stamped the American political sensibility since September 11, a cultivated fear that gave legitimacy, often carte blanche, to Bush's diplomatic and military decisions and a violent hubris that rested on overconfidence in the capacity of arms to protect democracy at home and extend it abroad. The self-proclaimed Hobbesianism of the neoconservatives vividly expresses this combination of the fear of death and pride in strength. America imagines itself a leviathan, a wounded leviathan, at once afraid to die and convinced of its invincibility. Political thought clearly must take stock of passions, not just interests, in trying to understand American foreign policy and its impact in turn on domestic politics and policies. Pierre Hassner, an especially astute and thoughtful analyst of international affairs, calls for a reflection on the "dialectic of the passions" in the world today, the dangerous interaction, for example, between Western fear and pride and Muslim humiliation and honor.[1] As a kind of contribution to that project, let us look more closely at the fear and pride woven into American policy in order to probe the symbols and institutions, practices and discourses, that shape, and are shaped by, these passions.

How could the combination of fear and hubris take hold within the public mind so persuasively, so effectively, after September 11? Donald E. Pease, Jr., has advanced an intriguing analysis under the title "The Global Homeland State: Bush's Biopolitical Settlement."[2] His analysis

---

[1]Pierre Hassner, "La revanche des passions," *Commentaire* 110 (Été 2005), pp. 299–312.
[2]Donald E. Pease, Jr., "The Global Homeland State: Bush's Biopolitical Settlement," Keynote Address, Conference on "Iraq and Its Consequences," sponsored by the U.S.-Europe

helps illuminate the symbolic-discursive patterns that have linked fear and hubris in the legitimation of administration policy. In the process Pease unravels the significance of the new reference to America as *homeland*. When the playwright David Mamet turned his finely tuned ear for American speech to this phrase, he remarked, "The phrase 'Homeland Security' . . . is confected and rings false, for America has many nicknames. . . . But none of us has *ever* referred to our country as The Homeland."[3]

As Pease points out, *homeland* has normally referred to the country that an American or his or her ancestors came *from,* the land left behind. The term *homeland* took hold after September 11, he surmises, insofar as it symbolized for Americans the sense of being newly alienated from the America they had always known, that is, from the place that had always been symbolized as removed, safe, innocent—the New World, the Virgin Land, the sea-protected continent of Manifest Destiny. When the trauma of September 11 tore away the illusion of American invulnerability and shattered those symbolizations, the Virgin Land was transformed into *ground zero,* and on this absence emerged the new symbolization *homeland* to redesignate America as the space Americans are displaced from and yearn for. Meanwhile, the USA Patriot Act and the new Homeland Security administration created new instruments of internal security that many feel deprive Americans of considerable civil liberties and liberal rights. Pease claims, in what is far more than a darkly ironic joke, that these measures made America a *homeland* in the exact sense of a place from which Americans are estranged, the place from which American rights and freedoms are missing.

What is to compensate for this loss? The official answer of course is that the curtailment of rights is compensated by increased security, but nothing could be more false, since the result has been a heightened sense of insecurity. Rather, Pease speculates, the people are compensated for their loss of rights and liberties at home by being offered the *spectacle* of the nation's military prowess as it deprives foreign peoples of *their* rights,

---

Seminar, Baruch College, New York, May 2004. I thank the author for providing me with a copy of the lecture. The essay that developed from it can be found in Donald E. Pease, Jr., "The Global Homeland State: Bush's Biopolitical Settlement," *boundary 2* 30, no. 3 (Fall 2003), pp. 1–18.
[3]David Mamet, "Secret Names," *Threepenny Review* 96 (Winter 2004), p. 6.

liberties, and sovereignty: from the overthrow of the Taliban and Saddam Hussein to the orange jumpsuits of Guantánamo.

The United States has put itself beyond the laws binding for other nations and at the same time created a *homeland* truncated of rights. These actions are best understood according to Pease in light of the idea of the "state of exception." Carl Schmitt's dictum that sovereignty is the power to declare the state of exception has been taken up by Giorgio Agamben and joined to the concepts of biopower and bare life. Pease draws on this framework to account for the suspension of rights found in the USA Patriot Act and the edicts behind Guantánamo and Abu Ghraib. The American citizen is reduced to bare life in the sense that he or she is merely a life-to-protect, as civic rights and civil liberties are shedded to whatever degree is necessitated by security measures, and foreigners are reduced to bare life in the sense that they are entities exposed to the fear of death by being stripped of legal standing and of the rights afforded by the U.S. Constitution or even the Geneva Accords. These two reductions to bare life are part and parcel of the state of exception, as the state, in particular the executive branch, abrogates to itself the unique power to say where law does and does not apply.[4]

Pease's interpretation incisively exposes a symbolic-discursive pattern (*ground zero, homeland, enemy combatants,* etc.) at work in the Bush administration's policies and vision of American global security. He tends, however, to construe this pattern as a structure that determines or is coextensive with the practices in question (that is, the Patriot Act, Guantánamo, the overthrow of the Taliban or Saddam Hussein, Abu Ghraib prison). Such an interpretation runs two risks. First, it overlooks the possibility that the practices might not work, or that they might work in a way that exposes the limitation or falsehood of their symbolic-discursive justification. Second, it overlooks the conflict within the political domain

---

[4]For the purposes of this discussion I follow Agamben's usage of the term "state of exception" to cover a wide range of situations that vary considerably in their treatment by Anglo-Saxon and Roman law as well as between various countries. State of emergency, martial law, emergency decree, suspension of habeas corpus, executive order, etc. all fall under Agamben's rubric "state of exception." Whether an umbrella term is justifiable and whether "state of exception" would be the most viable one are important points of discussion and debate, which Agamben himself engages in several places. For my purposes here, the question is moot since I am examining and challenging Agamben's association of the "state of exception" with the very definition of sovereignty and the very nature of the state.

over the symbols, discourses, and practices in question—conflicts that become all the more acute as the limitations and falsifications of the prevailing formation come to light.

Such political conflicts and such fissures between practice and justification afflicted the Bush administration's designs from the outset of the occupation of Iraq. Far from establishing a global homeland state, the United States failed to establish even basic security and civil order in Iraq. (So, too, in Afghanistan, where order hardly existed beyond the city of Kabul.) Abu Ghraib and Guantánamo were indeed spaces where the "state of exception" went horrifyingly far in reducing prisoners from rights-bearing persons to bare life. But the Abu Ghraib scandal prompted controversy, investigations, and disclosures that forced the Pentagon to change course in its handling of Iraqi prisoners. The Rumsfeld Pentagon was severely criticized in the press for fostering the climate of abuse and torture at Abu Ghraib. It had in effect transposed the (already questionable) techniques used to interrogate "enemy combatants" caught in the search for al Qaeda units in Afghanistan to the very different circumstances in Iraq, where the United States was an occupying power with the responsibility of the country's civil authority and where those rounded up were, when not common criminals or innocent bystanders, insurgents or Saddam Hussein loyalists. The public debate saw a distinction between the juridical-political spaces of Afghanistan and Iraq, between Guantánamo and Abu Ghraib. The space within which a state of exception could reduce prisoners to bare life, a space symbolized in the orange jumpsuits of Guantánamo, came up against a *limit* when it came to Iraq. The Abu Ghraib scandal redrew this limit and reaffirmed the distinction between an "enemy combatant" (a terrorist not acting on behalf of a state) and an insurgent in an occupied country.

A few weeks later the Supreme Court significantly curtailed the administration's handling of "enemy combatants" at Guantánamo and elsewhere by overturning the presidential authority to declare a U.S. citizen an enemy combatant, and it restored some minimal due process for aliens captured and held at offshore prisons. A more fundamental challenge to presidential power came in the Court's 2006 ruling in *Hamdan v. Rumsfeld,* which held that the Geneva Convention applied to "enemy combatants" and that Congress had to have a role in establishing procedures for any special tribunals. All these rulings contain ambiguities and limits and left much unsettled legally, and there is surely concern for how much leeway remains for the continued abuse of presidential authority. Moreover, Congress and Democrats remained fairly timid and acquiescent even after the Supreme Court in effect overturned the Bush administration's understanding of excutive power, largely because the public's commitment to rights and international law seemed far less intense than its craving for a sense of security and invincibility. Such an inclination was already obvious in the sorrowfully muted sense of scandal when controversy arose over the administration's "extraordinary rendition" of terrorist suspects into the hands of regimes known to torture and kill prisoners, a practice the president himself justified with unconvincing denials that torture was taking place and strong assertions of his duty to protect Americans.[5] In short, alarm over the extent and effect of the Bush administration's reliance on executive decrees and arbitrary power is thoroughly justified, but the brake that public opinion and court rulings have put on administration policies is a reminder that presidential power is not absolute and that even under what are widely perceived as wartime conditions, the balance of powers is capable of limiting presidential authority. The serious violations of the Constitution and international law that administration policy fostered came under political and judicial challenge.

Arendt versus Agamben

Such interruptions of the global homeland state forecast a protracted struggle to limit presidential power in the war against terrorism. They also put into question Agamben's theory and image of sovereign power. There is no room in Agamben's thought for the role of the division and balance of powers in defining the inner workings of modern democracy. He seeks, instead, to furnish the concepts by which the essence of

[5]Speaking at a televised news conference on April 28, 2005, President Bush replied to a question about "renditioning" by saying (my italics), "We operate within the law, and we send people to countries where they *say* they're not going to torture the people. But let me say something. The United States government has an obligation to protect the American people . . . . And we will do so within the law. And we will do so in honoring our comitment not to torture people. And we *expect* the countries where we send somebody to not to torture as well. But, you bet, when we find someone who might do harm to the American people, we will detain them and ask others from their country of origin to detain them. It makes sense."

modern state power as such can be understood and criticized. The sheer scope of such an ambition—along with the apparent reach of his fundamental concepts—makes his work all the more appealing. The tandem theorem of *biopower* and the *state of exception* deserves special scrutiny.[6]

Agamben draws inspiration for his understanding of "biopolitics" and "bare life" from a section of Hannah Arendt's *The Origins of Totalitarianism* devoted to the crisis in European politics and political thought that was produced by the masses of refugees set in flight during the wars, revolutions, and genocides of the twentieth century. Her discussion is provocatively titled "The Decline of the Nation-State and the End of the Rights of Man," and Agamben claims it as a precedent for his own theory. "Linking together the fates of the rights of man and of the nation-state," he writes, "her striking fomulation seems to imply the idea of an intimate and necessary connection between the two, though the author herself leaves the question open. The paradox from which Arendt departs is that the very figure who should have embodied the rights of man par excellence—the refugee—signals instead the concept's radical crisis."[7]

Arendt may inspire Agamben's project, but she does not provide a precedent for his argument. Her ability to "leave the question open" will stand in marked contrast to Agamben's mode of argument.

The twentieth-century refugee, according to Arendt, presented the Western world as never before with the need to act to protect the rights of man. Who more obviously than the *stateless* individual must enjoy inalienable human rights? "The Rights of Man, after all, had been defined as 'inalienable' because they were supposed to be independent of all governments." The history of the rights of man from their original declaration to the masses of modern refugees follows an intricate course. At the end of the eighteenth century, American and French

---

[6]The following discussion of Giorgio Agamben owes a great deal to many lively, sometimes contentious, always illuminating conversations with William McClellan, who first brought Agamben's work to my attention. His own engagement with Agamben's theory has been a benchmark for me. Cf. William McClellan, "'Ful Pale Face': Agamben's Biopolitical Theory and the Sovereign Subject in Chaucer's *Clerk's Tale*," *Exemplaria* 17, no. 1 (Spring 2005), pp. 103–34.

[7]Giorgio Agamben, "Biopolitics and the Rights of Man," in *Homo Sacer: Sovereign Power and Bare Life*, trans. Daniel Heller-Roazen (Stanford: Stanford University Press, 1998), p. 126.

revolutionaries effected the historic innovation of making "Man, and not God's command or the custom of history, . . . the source of Law." The rights of man were understood to precede and sustain the rights of the citizen. Moreover, the French Revolution even more than the American established the notion of popular sovereignty, of the "people" as sovereign. The relation of "man" and the "people" became central: "Man appeared as the only sovereign in matters of law as the people was proclaimed the only sovereign in matters of government." From this came a further implication: "it seemed only natural that the 'inalienable' rights of man would find their guarantee [in] and become an inalienable part of the right of the people to self-government. In other words, man had hardly appeared as a completely emancipated, completely isolated being who carried his dignity within himself, without reference to some larger encompassing order, when he disappeared again into a member of the people." Owing to the historical circumstances of the French Revolution and its aftermath in Europe, the "people" was conceived as a national people, a nationality. What had originally been declared to be every *individual's* inalienable *human* rights, preceding and sustaining the rights of the citizen, became instead "inextricably blended with the question of national emancipation: . . . it gradually became self-evident that the people, not the individual, was the image of man." When the system of nation-states, in the wake of their internal conflicts and wars with one another, began creating millions of refugees in the twentieth century, the refugees' *statelessness*—which should have evoked the very image of human rights—meant *nationlessness*, that is, the utter loss of any standing except being alive. The eighteenth-century philosophical innovation of abstract man's inalienable right had become for twentieth-century refugees the terrible reality of "the abstract nakedness of being human and nothing but human." Even countries founded on human rights did not react to protect the refugees' inalienable rights; nongovernmental international bodies proved ineffectual as states resisted interference in the name of national sovereignty; and refugees themselves sought not their human rights per se but national community: "The Russian refugees were only the first to insist on their nationality and to defend furiously against attempts to lump them together with other stateless people. Since then, not a single group of refugees or Displaced

Persons has failed to develop a fierce, violent group consciousness and to clamor for rights as—and only as—Poles or Jews or Germans, etc."[8]

The "abstract nakedness of being human" is of course what Agamben means by bare life. Nazi death camps revealed the biopolitical reduction or abstraction of human beings to bare life in its extremity. Arendt would have agreed. She might even have agreed when Agamben yokes this extremity with the Schmitt theorem: "if sovereign power is founded on the ability to decide on the state of exception, the camp is the structure in which the state of exception is permanently realized." The ambiguity contained in this statement marks where Agamben's mode of thinking parts company with Arendt's. Is he simply saying that the Nazi death camps amounted to a permanent state of exception or revealed what a permanent state of exception would devolve into? Or is he saying that the death camp reveals the *structure* of state sovereignty defined as the power to decide the exception? In the sort of slippage characteristic of his thought, Agamben is precisely saying the latter as though it follows logically from the former: "if the essence of the camp consists in the materialization of the state of exception and in the consequent creation of a space for naked life as such, we will then be facing a camp virtually every time that such a structure is created." The extremity of the death camp becomes a benchmark, even the epitome, of modern state power: "the birth of the camp in our time appears to be an event that marks in a decisive way the political space of modernity itself. . . . The camp . . . is the hidden matrix of the politics in which we still live, and we must learn to recognize it in all of its metamorphoses."[9] The metamorphoses of "the camp" in Agamben's account turn out to run the gamut from Auschwitz or the Vichy regime's roundup of Jews at the Vel d'hiver to the *temporary* detention of illegal Albanian immigrants in an Italian soccer stadium or of asylum seekers at French airports, all the way to housing projects and gated communities in the United States. Agamben thus furnishes his own argument with its reductio ad absurdum.

[8]Hannah Arendt, *The Origins of Totalitarianism* (New York: Harcourt Brace Jovanovich, 1973 [1951]), pp. 290–302.

[9]Giorgio Agamben, "What Is a Camp?" in *Means without End: Notes on Politics,* trans. Vincenzo Binetti and Cesare Casarino (Minneapolis: University of Minnesota Press, 2000), pp. 40, 42, 41, 44.

One sees here how misleading the claim to have described a *structure* can truly be. The Schmittian dictum on the state of exception is taken (dogmatically, axiomatically) as the unvarying, and complete, definition of sovereignty; the Nazis' devotion to establishing the legalities of states of emergency in orchestrating the Final Solution is taken as the exemplary instance of the role of law in the modern state; all manner of problems in modern society and politics—from spiritless wealthy suburbs to the Holocaust—are then taken as logical, quasi-inevitable consequences of the same structure of the modern state and sovereignty. Where Arendt tries to examine "the many perplexities inherent in the concept of human rights," Agamben transforms the dynamic contradictions she sees in the relations between man and citizen, human rights and civic rights, political identity and national identity, popular sovereignty and democratic self-rule, into the purely logical, fixed pairing of sovereign power and bare life.

Schmitt and Hobbes

The resulting theory of sovereignty rests on a particular imagination of power. Agamben constructs an *image* of power in which mass murder lurks in the rule of law, the nation-state inevitably churns out stateless multitudes, dehumanization is the consequence of declared universal human rights, the death camp is the paradigm of political modernity. The construct is built from bits of Schmitt and Hobbes. The drama of Agamben's thought lies there. It is worth examining how a thinker with such a sensitivity to modern catastrophes and an instinct for radical criticism of state power weaves that sensitivity and instinct together with the concepts of two thinkers who worshiped the state, Schmitt through his "political theology" and Hobbes in the form of his "mortall God." What emerges in Agamben is not just a theory of sovereignty and the modern state but also a philosophical-political sensibility that reveals a great deal about the aspiration to radical theory today.

The Schmitt theorem, *Sovereign is he who declares the state of exception,* sounds bold, but the aphoristic compression hides a more mundane line of reasoning: Sovereignty is the power (and responsibility, let us add) to maintain the rule of law; if the body politic is threatened to the point where to save it, including its rule of law, it is necessary to suspend fundamental laws, then it is the power (and responsibility) of the sovereign to declare this state of exception; if he failed to do so, his sovereignty (and perhaps the body politic) would end; if his declaration holds, his sovereignty is preserved and confirmed. Therefore, *sovereign is he who declares the state of exception.*

If one looks past the theorem's bracing rhetoric, the genuine political problem posed by the state of exception comes back into focus. Since the rule of law rests on the capacity to suspend the rule of law if necessary, whoever *declares* a state of exception will almost inevitably *claim* that it is necessary for the preservation of the rule of law and indeed of the body politic itself—no matter what aims and motives lie behind the decision. Whatever agency makes the declaration—an elected president, a prime minister's cabinet, an unelected head of state, a parliament, a monarch—lays itself open to challenge, limitation, even crisis. The very conditions of the legitimacy of its own rule shift. Legitimation may now take a dramatically simpler form but also a more precarious one, since it will likely depend more on immediate results than on institutionally or historically ingrained values. The state of exception undoubtedly enhances the opportunities for the abuse of political power and authority, but it does not inevitably unleash a downward spiral. Political systems can be resiliently self-correcting, especially as the public's sense of emergency wanes or the government's claim of necessity is thrown into doubt. In democratic states, disputes over this claim of necessity can erupt within the body politic or between branches of government. Contrary to the absolute categories purveyed by Schmitt, sovereign power does not transcend the potential clash inherent in the distinction between the *declaration* of the exception and the *claim* of its necessity.

The threshold at which a state of exception becomes necessary can only be determined by means of a judgment. No objective measure or legal principle suffices to discern when a peril to the state justifies declaring the exception. Leaders are undoubtedly tempted to exaggerate the peril in order to justify their declarations. But just as their decision itself is based on a judgment, so too their *appeal to necessity* opens the way for the public, the press, the legislature, and the courts to assume responsibility for scrutinizing and questioning the leader's decision and exercise a judgment of *their own.* Does the claim of necessity mask some other motive? Are valuable liberties being sacrificed to mere expediency? Does the perceived threat truly outweigh the value of the curtailed rights and procedures? Is the peril to the state as severe as claimed? Those sorts of questions are in play in every decision and debate over the state of exception.

Schmitt's "decisionism" needs to be deflated to recover what is really at issue, namely, that the extent of the peril facing the body politic and the means necessary to meet it can ultimately only be determined through *judgment* rather than logic, law, or knowledge. Agamben instead pumps more air into the decisionist balloon and presents the state of exception as a full-blown aporia: "But the extreme aporia against which the entire theory of the state of necessity ultimately runs aground concerns the very nature of necessity. . . . Not only does necessity ultimately come down to a decision, but that on which it decides is, in truth, something undecidable in fact and law."[10] There is no aporia here at all. The question before the body politic is "undecidable in fact and law" for the simple reason that it must be answered through a judgment in the political sense of the term. Judgment does not reduce to will-to-power or the purely arbitrary show of strength imagined by Schmitt. A judgment mixes, according to no fixed principle, attention to facts, concern for the law, opinions self-interested and disinterested, and persuasion.

Schmitt's absolutizing truism—*Sovereign is he who declares the state of exception*—has strangely gained in solemnity thanks to his association with Nazism. The Nazis came to power legally, secured their dictatorship by using the constitutional provision for emergency measures, and organized the death camps with careful attention to legalities. The recognition that dictatorship and even totalitarianism can arise from democratic institutions (and with popular support, let us add) underlies nearly all the most insightful historical and theoretical reflections on fascism. In the hands of writers like George Mosse or Claude Lefort such an awareness warns how fragile democratic institutions are, how dangerous the ineluctable role of the masses in modern politics is, and how rapidly all-embracing ideologies can transform social anxieties and communal loyalties into murderous passions. Agamben jumps to a simpler,

---

[10] Giorgio Agamben, "The State of Exception as a Paradigm of Government," in *State of Exception*, trans. Kevin Attell (Chicago: University of Chicago Press, 2005), pp. 29–30.

more radical conclusion: instead of reckoning with totalitarianism as (in Lefort's phrase) one of the "political forms of modern society," Agamben concludes that totalitarianism is the inherent, prevailing tendency of all political forms of modern society, that is, of the modern state as such.

In short, the totalitarian metamorphosis of the modern state in the Third Reich is taken to be the paradigm of the modern state. Such a gesture yields a language of radical political criticism, to be sure, but at the cost of ignoring or underestimating the same characteristics of democracy that Schmitt devalues and dismisses: diversity of opinion, separation of powers, plurality of interests. What is left obscured behind Schmitt and Agamben's dazzling theorem, *Sovereign is he who declares . . .* , is the little wedge created by the distinction—and hence the potential gap—between declaration and claim, act and justification, rule and legitimacy. Along these hairline fractures in the discourse of power lies the very possibility of a political realm and of democracy.

In his imagination of power Agamben also enlists Hobbes, not surprisingly of course, since *Leviathan* offers the first modern political theory of sovereignty. Agamben evokes it as foundational: "In the Hobbesian foundation of sovereignty, life in the state of nature is defined only by its being unconditionally exposed to a death threat (the limitless right of everybody over everything) and political life—that is, the life that unfolds under the protection of the Leviathan—is nothing but this very same life always exposed to a threat that now rests exclusively in the hands of the sovereign."[11]

What Agamben's summary leaves out is Hobbes's notion that the multitude, living in mortal fear in nature's (hypothetical) war of all against all, use their reason to make a covenant among themselves: they create their commonwealth by relinquishing their individual natural right to self-preserving violence and entrusting their safety and security to the sovereign power of their common body (the *common*wealth, the sovereign's body composed of the multitude's bodies). Yes, the natural right to self-preservation—to violence for the sake of self-preservation—is, so to speak, transferred to the sovereign. And, yes, Hobbes holds to a dark view of human nature and the necessity of constraints and threats

[11]Giorgio Agamben, "Form-of-Life," in *Means without End,* p. 5.

to rein in instinct. But one misses Hobbes's *political* theory altogether if one slides past his notion that the commonwealth-forming covenant lifts the coventers from the realm of natural right (where one may use any means necessary to preserve one's life) to the realm of natural law, that is, in his terms, from the state of nature to the laws of nature:

> For the Lawes of Nature (as *Justice, Equity, Modesty, Mercy,* and (in summe) *doing to others, as wee would be done to,*) of themselves, without the terrour of some Power, to cause them to be observed, are contrary to our naturall Passions, that carry us to Partiality, Pride, revenge, and the like. And covenants, without the Sword, are but Words, and of no strength to secure a man at all. Therefore notwithstanding the Lawes of Nature, (which every one hath then kept, when he has the will to keep them, when he can do it safely,) if there be no Power erected, or not great enough for our security; every man will and may lawfully rely on his own strength and art, for caution against all other men.[12]

The step from the state of nature to the laws of nature thus institutes in human affairs a new domain of values (justice, equity, modesty, etc.) that do not exist in the prepolitical domain.

There is another crucial aspect of *Leviathan* that Agamben does not acknowledge. This time, though, he tacitly accepts one of Hobbes's most questionable notions. The various forms of government—monarchy, aristocracy, and democracy in his vocabulary—are indistinguishable according to Hobbes when it comes to the nature of sovereignty. So undifferentiated is his conception of political power that he even insists that tyranny, oligarchy, and anarchy "are not the names of other Formes of Government" than monarchy, aristocracy, and democracy, "but of the same Formes misliked"! Thus, those who "are discontented under *Monarchy* call it *Tyranny;* and they that are displeased with *Aristocracy,* called it *Oligarchy:* So also, they which find themselves grieved under a *Democracy,* call it *Anarchy.*"[13] Writing almost a century and a half before the American founders based the Constitution on the division and separation of powers and the French revolutionaries introduced the universal

---

[12]Hobbes, *Leviathan,* pp. 223–224 [ch. 17].
[13]Ibid., p. 240 [ch. 19].

rights of man and of the citizen, Hobbes saw sovereign power as indivisible and best embodied in a monarch. By not distinguishing the sovereignty of monarchy, oligarchy, and democracy, he could argue that the monarch simply had the advantage of a more efficient, unified, and consistent rule, since sovereignty resided in but one person not several. Only by tacitly accepting this aspect of Hobbes's thought can Agamben hang onto the *image* of sovereign power as tyrannical, arbitrary, homicidal.

In sum, if (as Hobbes thought) sovereign power is in essence the same in all forms of state from democracy to tyranny, and if (contrary to what Hobbes thought) the covenant forming the commonwealth puts in the sovereign's hands a violence identical to the war of all against all, then the body politic reduces to Agamben's triptych of bare life, mortal fear, deadly force. And if (as Schmitt thought) Germany under Hitler openly enacted the hidden essence and true logic of sovereignty, then the modern state in all its forms embodies just this biopolitical monster. Agamben thus arrives at the core concepts of his radical criticism of modern states' sovereignty.

While Donald Pease shows how productive those concepts can be in exposing the dire designs of government policies, I do not accept Agamben's logic of power and image of power. There is to be sure no difference between torture authorized by the Pentagon and torture when practiced by fascists. The violation of human dignity through humiliations and threats designed to induce in the victim the fear of being killed is always the same, no matter what acts the victim has committed or what ideals and motives the torturer holds. The chill and outrage that come from feeling Guantánamo's resemblance to the interrogation chambers of fascist regimes must nevertheless not obscure the fact that the gap between the *declaration of the exception* and the *claim of its necessity* is a space where genuine political and juridical dispute take place. It is wiser to seek the paradigm of the "political space of modernity itself" in such spaces of contestation rather than in the prison camp at Guantánamo.

Decision and Covenant

Quarrels in political theory bear ultimately on the resources of political criticism. At stake in today's controversies are the language, concepts, and modes of argument that might send a few beams of light through the fog into which terrorism and war have plunged democracy. Agamben's mode of argument is provocative in large part because, reflecting a recent trend in political commentary, it extracts certain concepts from thinkers associated with the extreme right and uses them for ostensibly progressive or radical democratic purposes. Against this trend it is often remarked, usually in histrionic tones of astonishment, that any philosophical claim made on the basis of concepts from a devoted Nazi like Carl Schmitt at best is tainted and at worst perpetuates fascist thinking. Such an attitude is woefully ill-founded, and it misses the mark when it comes to the intriguing question of why a thinker like Schmitt has become a point of reference for leftist thinkers.

The answer lies in the vicissitudes of Western Marxism, which came of age and thrived during the Cold War and then succumbed to crisis in its reactions to the fall of communism in 1989. The key to its crisis lay in its response to the Cold War divide of Soviet totalitarianism and Western capitalism. Why did Marxism have the ground cut from beneath it with collapse of Soviet communism even though it had lent little or no support to the Soviet Union? All during the Cold War, Western Marxism forged a two-pronged discourse in search of a socialist vision that repudiated both American-led capitalism and Soviet-dominated totalitarianism. One prong criticized capitalism and the excesses of Western anticommunism lying behind repressive domestic policies (from McCarthyism in the 1950s to West Germany's antiterrorist campaign in the late '70s) and neo-imperialist foreign policies (Vietnam, Chile, Nicaragua). Democracy per se was simply taken for granted, all the more so because of its reassuring stability and security in most Western countries, while the anticommunist excesses were blamed for inhibiting the creation of a more egalitarian society. The other prong denounced state socialism and imagined that every revolt in the Soviet bloc (Hungary, Czechsolvakia, Poland) was, despite its inevitable brutal repression by the Soviets, the harbinger of the eventual transformation of state socialism into something democratic. The two prongs complemented one another, as mutual alibis, so to speak: anti-anticommunism presupposed democracy while rejecting capitalism; antitotalitarianism presupposed socialism while rejecting bureaucracy and one-party rule. The rejection of capitalism, bureaucracy, and one-party rule seemed then to confirm the presupposition that socialism and democracy belong together. Meanwhile, the standoff of the Cold War itself deferred the crucial unanswered question: by what path could liberal democracy become socialist or state socialism democratic? When the Soviet system collapsed and the Cold War ended, Western Marxism had to face two uncomfortable truths: state socialism had never been reformable, and democracy has no intrinsic affinity with socialism or even social justice conceived in egalitarian terms.

New intellectual models were sought to link anew democracy and socialism, anticapitalism and expanded rights. While Agamben has essentially discarded the Marxist paradigm, he remains loyal to the aspiration that Marx gave to theory, namely, according to his oft-cited dictum, to be *radical* in the etymological sense of getting to the *root*, that is, the root of society, of politics, of history. How to live up to this aspiration after having long ago abandoned, or never held, the idea that the struggle between classes or the labor theory of value was the explanatory root of society, politics, and history? In particular, if the state does not arise, as Marx thought, from the economic surplus controlled by the ruling class, what then is the *root* of the state? Enter Carl Schmitt. The sovereignty theorem is nothing less than a substitute root-thesis. The modern state—at bottom, at its origin, at heart, at the root—rests on arbitrary decision and violence. Schmitt celebrates decisionism, while Agamben turns it into the perfect tool for denouncing whatever aspect or action one might want to criticize in the modern state by tracing it implacably, logically, back to the root in violent arbitrary will.

My objection to Agamben's appropriation of Schmitt can be amplified via two Arendtian themes. The first is the distinction between *power* and *might*, and the second is the notion that the inauguration of the political realm—polis, polity, state—is *ungrounded*. Schmitt's concept of the political collapses the distinction between power and might; indeed, when it comes to the modern state, Agamben too slides synonymously from power to strength, might, violence. As a result, every form of legal, rule-governed power in the modern state acquires the aura of arbitrary violence. His recourse to Schmitt's decisionism is the source (or perhaps, conversely, the effect) of this equation of power and might. Schmittian decisionism plays another role as well. It provides an image (or myth or metaphor) of the step from the prepolitical world to the political. The problem is not the recourse to image, myth, or metaphor. All political theory must do just that when it seeks to think the inauguration of the political realm. The problem is that Agamben chooses *this* myth. The decisionist myth postulates the origin of sovereignty without a covenant (contrary to Hobbes) and therefore imagines power as indistinct from sheer might.

Eschewing the distinction of power and might, the Schmittian myth construes the founding of the political realm as an act of dominion. By following suit, Agamben short-circuits a crucial theoretical question: if the founding of the political realm, if the inauguration of a body politic, is *arbitrary*, what is the nature of this arbitrariness? For it is hardly self-evident philosophically or historically that the arbitrary founding act is necessarily a matter of dominion and violence. The political realm is indeed ungrounded, in that it does not issue from any divine, metaphysical, or natural principle. It results from a human *decision*, but Schmitt limits the decision to but a single image: one entity overpowers others, making itself sovereign by virtue of its strength and exempting itself in principle from the rules imposed on those it rules over.

When Arendt ponders how the political realm arises from the fabric of prepolitical relations (clans or tribes), she normally reflects on the ancient Greek polis but also draws on another example because of its specific relevance to the political ideas that would eventually forge American democracy. The Mayflower Compact was drawn up by the first Puritan colonists just before they disembarked at Cape Cod, several hundred miles off course from their original destination of Virginia. It is 1620, some two decades before Thomas Hobbes begins publishing. The colonists' inaugural gesture is a covenant:

> We whose names are underwritten . . . do . . . solemnly and mutually in the presence of God and one of another, Covenant and Combine ourselves together into a Civil Body Politic, for our better ordering and preservation and furtherance of the ends aforesaid; and by virtuue hereof to enact, constitute and frame such just and equal Lawes, Ordinances, Acts, Constitutions and Offices, from time to time, as shall be thought most meet and convenient for the general good of the Colony, unto which we promise all due submission and obedience.

What Arendt finds remarkable is that the Puritans' interest in covenant as they encountered it in the Old Testament cannot explain their *political* action on arriving in the New World. "For the Biblical covenant as the Puritans understood it was a compact between God and Israel by virtue of which God gave the law and Israel consented to keep it, and while this covenant implied government by consent, it implied by no means a political body in which rulers and ruled would be equal, that is, where actually the whole principle of rulership no longer applied."[14] The Athenian resonance of the idea of being equals in ruling and being ruled does not mean that the Greek polis served as precedent for the Pilgrims; on the contrary, Arendt argues, "Nothing but the simple and obvious insight into the elementary structure of joint enterprise as such . . . caused these men to become obsessed with the notion of compact and prompted them again and again 'to promise and bind' themselves to one another."[15]

According to William Bradford's account in *Of Plymouth Plantation* (book 2, chap. 9), the Compact was made in response to the immediate uncertainties of the situation. There were restive voices on board, including "the discontented and mutinous speeches" of some of the non-Puritans. As things stood among the colonists, "none had power to command them, the patent they had being for Virginia and not New England, which belonged to another government, with which the Virginia Company had nothing to do." Having no ground to stand on, the Pilgrims decided that the Compact—this "combination," in Bradford's words, made by them as "the first foundation of their government"— "might be as firm as any patent." Their compact *is* the arbitrary unpredetermined ungrounded *decision* they took. Schmitt's decisionism cannot comprehend this inaugurating decision. Of the motifs in the Mayflower Compact that anticipate Hobbes, Schmitt's imagination of power can incorporate protection but not covenant. "Action in concert" (Arendt) falls outside what the Schmittian imagination will entertain. "The *protego ergo obligo* is," he asserts, "the *cogito ergo sum* of the state."[16] But, a Schmittian might well ask, doesn't the Mayflower Compact itself affirm obedience as well as protection? Indeed, it does: "we promise all due submission and obedience." *Submission* and *obedience* are part of

[14]Hannah Arendt, *On Revolution* (New York: Penguin, 1990 [1965]), p. 172.
[15]Ibid., p. 173.
[16]Carl Schmitt, *The Concept of the Political*, trans. George Schwab (Chicago: University of Chicago Press, 1996), p. 52.

the Schmittian lexicon, but the phrase also contains two words that Schmitt ignores or bans in his theory of political foundations: "we *promise* all *due* submission and obedience." *Promise* designates the covenanters' nonviolent founding act, and *due* confirms that the rules to which the covenanters relate, equally as rulers and ruled, must be "just and equal Lawes" designed for the "general good" if they are to command obedience. In the Arendtian myth of political inauguration, action in concert initiates the founding of government, whose laws (*lege*) are legitimate only insofar as they are just and equal.

Arendt adheres to the tradition of civic democracy going back to the ancients, whose political thought she excavated and reinterpreted through the modern temper. Her commitment is often criticized as anachronistic and one-sided in light of the more modern values associated with liberal democracy and social democracy. The tradition's archetypes do indeed reek of anachronism: virile Greeks mill about the marketplace readying the arguments that will shape the destiny of their polis; Florentine gentlemen renew their efforts at self-government while puzzling over the ancient authority of Aristotle and Cicero and reading divine revelation and grace in Charles VIII's military expedition to Italy; New England farmers, tradesman, and merchants gather in town meetings to levy taxes and repair roads; America's stylish slaveowners and homespun patriots rub elbows, contemplating themselves as equals, as they sign their names to the Declaration of Independence or the Preamble to the Constitution. Anachronistic though they may be, these are the primal scenes of democratic citizenship that dot the history of Western political thought.

Several meanings were folded into the ancient idea of the *polis*. It meant the polity (the political community of citizens), the public realm (the space where citizens gather to discuss and persuade, compete and decide), the *res publica* (the matters of public concern), and most tangibly it meant civic life and therefore ultimately the city itself. Athens was at once dwelling, fortress, and goddess. Moderns have long looked in awe at the ancients' way of experiencing their city and world. From Schiller and Blake to Nietzsche and twentieth-century historians and philosophers, Western thought has probed and criticized our modernity by trying to imagine how the ancients imagined things. Blake, anticipating Nietzsche, castigated modern religion, morality, and rationalism itself as deadening inversions of the poetic spirit by which the ancients imbued their world with the sacred. Athena was not a goddess separate from the city but the city itself experienced *poetically* in its divinity: "The ancient Poets animated all sensible objects with Gods or Geniuses, calling them by the names and adorning them with the properties of woods, rivers, mountains, lakes, cities, nations, and whatever their enlarged & numerous senses could perceive."

Whatever the Greeks' experience of the sacred aura of their democracy may have been, the multiple meanings of the polis—body politic, publicness, commonweal, city, and civility—were an integral whole. By contrast, those meanings have dispersed into fragmented, often conflicting notions in modern democracies. The *city* and the *citizen* have become nation and electorate; the *statesman-demagogue* has become the professional politician at the head of a mass party; the *citizen-soldier* has been replaced by military professionals, the arms industry, and weapons of mass destruction; the *common culture,* which the Athenians experienced in their religious rites, tragedy festivals, and philosophic banquets, has been replaced by mass culture and consumerism; and the sacred *dwelling* has burgeoned into the disenchanted metropolis with its luxury condos and housing projects, malls and ghettos, corporate headquarters and suburban sprawl.

Why then even bother with the ideas and symbols of civic democracy or, as it has been variously called, republicanism, civic humanism, the "Machiavellian moment"? The civic democratic tradition remains relevant first and foremost for a negative reason: without it modern democracy lacks any conception of active citizenship. Liberalism foregrounds individual liberty as freedom from the state, and social democracy foregrounds the state's obligations to the citizen. Only civic democracy concerns itself with the meanings and values of belonging to a political community and the rules and practices of political participation.

In reflecting on the paradoxes and dilemmas of the modern nation-state, Arendt is principally concerned with the preservation of citizens' rights and participation because she believes that participation is the only true countermovement to the atrophy or collapse of the permanently fragile polis. Her analysis of refugees and human rights thus stressed the value of participation over belonging, the individual over the people, self-rule over nationhood, the citizen over "man." And yet she realizes that the values she asserts do not, and cannot, simply dispel belonging, popular sovereignty, nationalism, or "man." In my view, Arendt hesitates to make these distinctions absolute—stays true to her habit and instinct of keeping the question open—because she tacitly realizes that while *belonging* and *participation* are the twin determinations of citizenship, they come into tension, often contradiction, in *modern* democracy. Ancient citizens enjoyed their equality with one another on the social foundation of their unequal relations with others, specifically the women and slaves of their households, strangers, and the city's other nonslaveholding inhabitants, including artisans. Participation went hand-in-hand with the sense of belonging to a secure social group. Modern democracy undoes the social cohesiveness among citizens by extending citizenship in principle to all social groups.

The French Revolution introduced universalism into the public realm. It did so by giving all members of society a claim to the rights and freedoms of citizenship. This universalism was an *event,* not a completed but an inaugurating event, that is, an event inaugurating the possibility of—. No longer could the social conditions of *belonging* anchor and justify the political rights of *participation.* It took nearly two centuries for the restriction of citizenship according to property, gender, and race to be overcome in Western democracies—thanks to successive upheavals and struggles, whose watersheds include the British Reform Bills, innumerable battles of the European, British, and American labor movements, the Fourteenth Amendment, women's suffrage, and the Voting Rights Act. As the rights of political participation become universal, the social cohesiveness among citizens is volatilized. On the one hand, social inequalities become a source of conflicts *within* the political sphere rather than, as with the Athenians, the relatively stable realm *from which* the political sphere distinguished itself. On the other hand, the forms of *belonging* to the political community become volatile as citizenship is associated with nationhood and popular legitimacy.

The Ordeal of Universalism

Arendt draws from the civic democratic tradition those values of self-rule that political modernity puts at risk. Hence her emphasis on power as action in concert and political inauguration as mutual pledging.

Politics for Arendt is principally an arena of judgment and persusasion: "The judging person—as Kant says quite beautifully—can only 'woo the consent of everyone else' in the hope of coming to an agreement with him eventually. This 'wooing' or persuading corresponds closely to what the Greeks called *peithein,* the convincing and persuading speech which they regarded as the typically political form of people talking with one another."[17] How then do things stand with judgment and persuasion in the context of political modernity?

The modern crux for thinking through the nature of judgment lies in Kant's *Critique of Judgment.* Not just aesthetic judgment but also political judgment. Aesthetic judgment turns on the claim *this is beautiful;* political judgment on the claim *this is unjust.* Even though the two forms of judgment diverge at their root and are often antagonistic in their aim, they are linked because publicness is their shared condition of possibility. Political and aesthetic judgment both arise from within the public realm. The beautiful, according to one of Kant's central arguments, "gives pleasure with a claim for the agreement of everyone else." What does "*with* a claim for . . . agreement" mean? Along with? accompanied by? incidentally combined with? No, the thesis is stronger: the experience *this is beautiful* is inseparable from an appeal to others that they too find this beautiful. More strongly yet, the experience of the beautiful *happens as* a claim for the assent of others. My experience of beauty tacitly carries within it this appeal to others; conversely, only insofar as I tacitly appeal to others that *this is beautiful* do I *experience* beauty. In Kant's words, "By this the mind is made conscious of a certain ennoblement and elevation above the mere sensibility to pleasure received through the sense, and the worth of others is estimated in accordance with a like maxim of their judgment."[18]

Art makes its appearance in the public realm, and what Kant takes account of in a new way are the implications of this publicness for the individual's inner experience of apprehending an artwork. Since the beautiful "gives pleasure with a claim for the agreement of everyone else," my own participation in the *public* realm is integral to my *inner* experience of the artwork and beauty. "Taste judgments," as Arendt puts it, "share with political opinions that they are persuasive."[19] We owe to Arendt the realization that Kant's reflection on the tie between judgment and publicness in aesthetic experience is at the same time "perhaps the greatest and most original aspect of [his] political philosophy," because he in effect "classif[ies] taste ... among man's political abilities"[20] and discovers in the maxims that epitomize taste or aesthetic judgment those on which political judgment hinges, especially the maxim of enlarged thought or the enlarged mentality: "to put ourselves in thought in the place of everyone else."

Arendt reaches behind Kant to the ancient polis, and more specifically in this instance to the Romans, for a landmark concerning the tie between aesthetic and political judgment. The concept of culture for the Romans had the threefold significance of "developing nature into a dwelling place for a people," "taking care of the monuments of the past," and what Cicero called *cultura animi,* which, Arendt says, "is suggestive of something like taste and, generally, sensitivity to beauty." Once a work is produced and appears within the space of the public realm, "every single judging person" (Kant) can judge its value as part of the human dwelling, as a monument soliciting care and preservation, as something of beauty. This judgment evinces a political faculty because it intrinsically involves an enlarged mentality and persuasion. What judgment appeals to is the *sensus communis*—common sense as "community sense"—which, according to Arendt, "discloses to us the nature of the world insofar as it is a common world. . . . Judging is one, if not the most, important activity in which this sharing-the-world-with-others comes to pass. . . . The activity of taste decides how this world, independent of its utility and our vital interests in it, is to sound and look, what men will see and what they will hear in it."[21] The judging and sustaining of this common world is, then, the aesthetic-political foundation of culture as dwelling, monument, and beauty.

Judgment appeals to the sensus communis, that is, the shared sense of an actual community in the midst of its common world because taste is always open to dispute. "Hence judgment," Arendt contends, "is

[19]Arendt, "The Crisis in Culture," p. 222.
[20]Ibid., pp. 219, 223.
[21]Ibid., pp. 221–22.

endowed with a certain specific validity but is never universally valid. Its claim to validity can never extend further than the others in whose place the judging person has put himself for his considerations."[22] Now, such a conception holds good for the Greek polis because its citizens' equality with one another rested on their social homogeneity, but political modernity volatilizes the sensus communis of the political community along with the social cohesiveness among citizens. The universalism that the French Revolution thrust into the public realm was, as I have said, an inaugurating event. Just as the founding principle of the American republic—*All men are created equal*—troubled the slaveholding reality, so the French declaration of the rights of man and citizen was not realized, arguably is not realizable, but its universalism recurrently renews itself as a source of political criticism, struggle, and innovation.

I will illustrate my view with a passage from James Baldwin's *The Fire Next Time*, written in 1962 in the midst of the agony and hope of the Civil Rights movement. The passage says more about the critical power of universalism than our political philosophy could dream of. The context is Baldwin's reflection on a kind of paranoia afflicting the everyday experience of blacks in America:

> it begins to be almost impossible to distinguish a real from a fancied injury. One can very quickly cease to attempt this distinction, and, what is worse, one usually ceases to attempt it without realizing that one has done so. All doormen, for example, and all policemen have by now, for me, become exactly the same, and my style with them is designed simply to intimidate them before they can intimidate me. No doubt I am guilty of some injustice here, but it is irreducible, since I cannot risk assuming that the humanity of these people is more real to them than their uniforms. Most Negroes cannot risk assuming that the humanity of white people is more real to them than their color.

With that last sentence—"Most Negroes cannot risk assuming that the humanity of white people is more real to them than *their color,*" that is, their own color, their whiteness—Baldwin touches the core of racism. It is not only, or fundamentally, white people's dehumanization of blacks

[22]Ibid., p. 221.

but their dehumanization of themselves: they value their race over their own humanity.

The universalism evoked here in valuing humanity is not the abstract universalism so widely repudiated in the various discourses of antiuniversalism and antihumanism; rather it is a value that arises in Baldwin's everyday experience, as it did in the Civil Rights movement itself, as an agonized challenge to whites' disavowal of their own humanity in their enjoyment of color and power. The disciplined, confrontational nonviolence of the Civil Rights movement in turn overcame the paranoia Baldwin so movingly documents. The nonviolent protesters dared to risk assuming that whites' humanity ultimately did count more to them than their color and power. The movement at that moment undertook, indeed led the entire nation into, the *ordeal* of universalism.[23]

In this sense, universalism is never a given or an achievement. Existentially and politically, it is an ordeal. In the political discourse and judgment inaugurating and sustaining such an ordeal, the universal expresses itself negatively. It turns on the claim *this is unjust* and appeals to the "worth of others" in calling on them to recognize the injustice.

Keeping in mind this understanding of how the democratic revolution inaugurated two hundred years ago affects political judgment, let us return to aesthetic judgment and try to assess how it changed with the advent of political modernity. Empirically, every community of taste has particular habits, competences, preferences, accepted conventions, and standards; indeed, those are precisely the constituents of its sensus communis. But in a modern democracy no single community of taste coincides with the polity as such—as was the case for the ancients *in fact* and for Kant *in principle*. In the Arendtian-Kantian perspective, every individual engaged in aesthetic judgment reaches beyond him- or herself, putting oneself in the place of everyone else, by drawing on the intuition that the horizon of everyone else's judgment is the same sensus communis that has shaped one's own sensibility. Modern democracy, however, multiplies such horizons and so fractures or breaches the

[23]Nonviolent civil disobedience is, at the other pole from emergency decrees, another form of the "state of exception," in the precise sense of a temporary, disciplined crossing of the boundary of lawfulness for the purpose of achieving something *on behalf of* the rule of law that cannot, in the judgment of those taking action, be accomplished by merely obeying the law.

boundary of every actual community of taste. There are today multiple communities of taste within a single body politic; their complementarities and antagonisms, their overlaps and differences, give contemporary criticism and cultural debate its verve and much of its confusion.

Many are the ways to escape the conundrum of plurality and universalism. Liberals tend to retreat within the borders of an existing community of taste and declare that its particular values *are* universal. Conservatives make a similar retreat and assert, with more or less aristocratic overtones, that they are in firm possession of established standards sanctioned by tradition by which to make aesthetic judgments; this is why Hilton Kramer calls his journal *The New Criterion*—and why, of course, T. S. Eliot called his *Criterion.*

Against the conservatives, I return to one of the most radical moments in Kant's aesthetic thinking. For Kant did not believe that the beautiful brought to appearance in the artwork conformed to any existing standard. On the contrary, he distinguished aesthetic judgment from other modes of thinking because in it the particular is not derived from a rule, but rather the rule has to be derived from the particular. The beautiful exists only in the particular. It is fortuitous, unexpected, unforeseen, unprecedented. Milan Kundera touches on other implications of just this when he provocatively asserts that "the history of an art is a revenge by man against the impersonality of the history of humanity."[24] Philippe Sollers hits a similar nerve when he says that every artwork is something that "should not have existed."[25]

Against the liberals, it is necessary to take Kant one step further. For what is the "rule" derived from the particular in aesthetic judgment? And in what sense is it universal? The rule is nothing other than the *aim* of my appeal for the agreement of others, which can only be affirmed by the response and agreement of (all the) others. However—or, rather, *therefore*—it is unattainable. Kant thought of this as a merely empirical circumstance: even as my judgment must *in principle* appeal for the agreement of everyone, I know that *in fact* not everyone will agree. The situation today is, instead: I know *in principle* that not everyone will

[24]Milan Kundera, *Testaments Betrayed: An Essay in Nine Parts,* trans. Linda Asher (New York: HarperCollins, 1995), p. 16.
[25]Philippe Sollers, *Théorie des Exceptions* (Paris: Gallimard, 1986), p. 12.

agree, but my judgment must *in fact* appeal for the agreement of everyone. The reason for this is doubly negative. The plurality of communities of taste negates the expectation of universal agreement, but this same plurality also negates any community's claim to possess, solely within the boundaries of its own sensus communis, a universally valid judgment. Universalism in our time is the work of the negative, of this two-pronged negation. The universal is ever beyond our grasp, yet it must always be the aim of our reach. Therein lies the aesthetic ordeal of universalism.

The philosopher Gianni Vattimo has aphoristically captured something of the consequence of this new aesthetic-political situation: "To live in this pluralistic world means to experience freedom as a continual oscillation between belonging and disorientation."[26] In my view, this experience also fuels the temptation to abandon the universal, that is, to relinquish the always troubled reaching beyond the bounds of existing community. That temptation, in its universalist and antiuniversalist forms, must be resisted. It is necessary, instead, to embrace the truth of relativism *and* the ordeal of universalism.

[26]Gianni Vattimo, *The Transparent Society*, trans. David Webb (Baltmore: The Johns Hopkins University Press, 1992), p. 10.

Thus we call a belief an illusion when a
wish-fulfillment is a prominent factor
in its motivation.   —SIGMUND FREUD

SEPTEMBER 11 AND
FABLES OF THE LEFT

First Response

On September 11, 2001, thousands of our fellow city-dwellers van-
ished, spectacularly and invisibly, before our eyes. Whether seen against
the indifferent blue of that morning's stunningly beautiful sky or on television, the devastation of the World Trade Center towers overawed wit-
nesses in the city and the world. Americans felt the shock of realizing as
never before that our civic life is fragile and our global power dangerous.
The attacks also called for unprecedented, difficult political judgments.

The Bush administration's decision to undertake a concerted military and diplomatic offensive in response to the massive attack on
American soil forced it onto unforeseen and unwelcome terrain. The
Republicans assumed the presidency loaded with three guiding ideas on
foreign policy—isolationism, unilateralism, and the Powell doctrine—
all of which had to be suspended in the hours and days after the attacks.
How significantly those prejudices will be revamped as a result of the
war on terrorism remains to be seen. The war in Afghanistan in effect
confirmed rather than reversed the administration's prejudices, as the
Taliban regime collapsed and al Qaeda was routed more quickly than
the public and perhaps the Pentagon anticipated. The Republicans remained isolationist when it comes to genocide and "nation-building"
and unilateralist when it comes to global warming and missile treaties,
and the Powell doctrine remained deeply ingrained in military thinking. Nevertheless, the administration embarked on an altogether different kind of military and diplomatic offensive in the face of a potent but

stateless enemy. And the fact that the dispersal of al Qaeda did not reverse the growth of Islamic radicalism meant that the nation and the administration faced susbstantially the same challenge after the fall of the Taliban as they did on September 11.

As for the Democrats in Congress, they faced an all too familiar challenge, one they had rehearsed in December 2000 when they declined to risk a political, let alone a constitutional, crisis and capitulated to the Supreme Court's giving George W. Bush the presidency without our knowing who had actually been chosen by the voters. The Democrats once again put national harmony above party advantage—or, from a less generous point of view, put the potential benefits of power sharing above principle and debate, ultimately at the cost of their future ability to offer the voters an alternative to the Republicans. As regards the conduct of the war on terrorism itself, it is doubtful that they had an independent view, since the Clinton years had left Democrats without an articulate vision of foreign policy. Beginning with their acquiescence on the USA Patriot Act, they even evacuated the terrain of civil liberties and immigrants' rights that has differentiated them for two decades from Republicans and the Reagan-appointed judiciary.

In sum, the Bush administration initially responded to September 11 with realism and inventiveness, while the Democrats brooded quietly and uncertainly, hoping eventually to hatch some agenda of their own. At the outset of the war on terrorism neither Republicans nor Democrats, the one by design, the other by passivity, have honored the ideal that Pericles articulated for a democracy at war: "we do not consider discussion an enemy of dispatch; our fear is to adopt policy without prior debate."

Those of us who generally hold a highly critical attitude toward American foreign policy and the direction of civic democracy and social justice in recent decades were also ill-prepared by our political preoccupations and intellectual habits to respond to September 11. If I identify "we" as the Left or the American Left, readers will presumably know what I mean. The term, though, is wearing thin, first, because the cohesiveness of the movements, activists, and intellectuals referred to is rather more imaginary than effective, and, second, because the belief that there *is* a Left has in recent years served as little more than a palliative providing a sense of belonging. September 11 posed a stark question: Should the United States mount a military and diplomatic offensive against terrorist networks, the Islamic fundamentalist organizations that spawn them, and the regimes that harbor them? To me, the answer has been emphatically Yes. The agenda and stakes of political commitment changed on September 11. In politics one seldom gets to choose one's battles, and the slogan that politics is the art of the possible, which is typically evoked to express self-assured realism or apologetic pragmatism, in fact contains the darker recognition that knowledge and decision in politics are ineluctably belated.

Many voices on the left spoke out in the immediate aftermath of the attack with a profound lack of political judgment, often compounded by a seeming lack of compassion and horror. The apparent lack of compassion and horror was, I believe, merely apparent. I do not imagine that these writers were affected less than anyone else by the destruction and death. Many comments suggested a failure of words, a simple, though perhaps telling, inability to express terror and pity in the same breath as dissent and criticism. However ambiguous or clumsy their failure of words may have been, the political judgments at issue are another matter.

Opponents of the American war on terrorism variously advanced essentially four arguments:

1. Armed force was an unnecessary and excessive reaction to the September 11 attack.
2. An international police action, focused on the United Nations and international courts addressing crimes against humanity, would have been a more appropriate and/or more effective mechanism for responding to September 11 than a U.S.-led military and diplomatic offensive.
3. U.S. actions past and present are the true cause of the terrorist attack, and, therefore, addressing the grievances in the Arab and Islamic world is the most appropriate (or only justifiable) course of action to take.
4. The source of the rise of Islamic fundamentalism and terrorist organizations is ultimately the Palestinian-Israeli conflict, and the United States should therefore impose a solution on that conflict, including the establishment of a Palestinian state, rather than pursuing a military offensive against al Qaeda.

The argument that the United States should show complete restraint and avoid military action, seeking instead a "peaceful response" or "nonviolent solution" in the phrasing of several petitions that circulated, might have carried weight if backed up by the conviction that the September 11 attack represented a kind of culminating achievement or last hurrah of the Islamic terrorist organizations. But there was every reason to believe the contrary; the terrorists revealed their willingness and capacity to escalate their assault on the United States, other nations, and the world economy. Alternatively, one could have argued that the United States and other countries have little alternative except to absorb the blows to their people and economies until *some* other solution can be found. But no one I have read and none of the petitions I saw after September 11 straightforwardly claimed that the threat was over or that the possible harm to come was worth absorbing. In the absence of those arguments, the peace activists advanced wish-fulfillments in the guise of political principles.

Multilateral Ambivalence

Others argued that instead of a war against terrorism a police action should be pursued against those who perpetrated the September 11 attack. This argument sometimes includes the claim that the attack was a crime against humanity rather than an act of war and advocates using international courts to prosecute the criminals. There is a crucial difference between the situation faced on September 11 and the Nuremberg trials or the prosecution of atrocities in Bosnia; those war crimes trials occurred after the cessation of hostilities and the capture of the criminals. After September 11, the Islamic terrorists were still active, their planning of violent acts ongoing, and their leaders "at large." A police action also presupposes that a policing power holds a relatively effective monopoly on legitimate violence in the territory where the criminals operate and hide. Afghanistan was not such a place. Had the Taliban rulers of Afghanistan arrested Osama bin Laden and his network, then indeed the international juridical apparatus, even Afghan courts, could have played a role. Without that, the pursuit of al Qaeda inside Afghanistan required a level of organized force more akin to war than policing.

On the face of it, the United States simply opted for multilateral

diplomacy and the unilateral use of force. Such a stance represents a refusal to subordinate national prerogatives to the decisions of any international body. Even though the Bush administration has demonstrated,
before and since, a dangerous disdain for the principle of international
decision making via the United Nations, I think they were right in this
case. Moreover, the U.S. position is not as unilateral as it may seem. The
launching of the war on terrorism, up through the actual campaign in
Afghanistan, amounted to the United States acting in self-defense *and*
in the interests of many other nations in Europe, the Middle East,
and Asia.

There are deep-seated ambivalences within the international community when it comes to the shadings of unilateralism and multilateralism on the part of the United States. Criticisms of American will-
fulness are frequently counterbalanced by the expectation, not simple
resignation, that the United States will act diplomatically and militarily
in its own interests. In the wake of September 11, the nations on the Security Council neither advocated a UN-led offensive against al Qaeda in
Afghanistan nor opposed a U.S.-led military action. Indeed, it is not
clear that the UN is prepared to undertake such an offensive. While it is
indeed true that the UN's capacity for action has been decisively inhibited by the antagonistic and arrogant positions of the United States itself for several years, largely because of the anti-UN ideology of Senate
Republicans, its unreadiness to take charge of the war against terrorism
was a considerable part of the present situation. Moreover, the complex
diplomatic imperatives of the UN itself often do not easily square with
the necessity for concerted military action against terrorist networks,
which are often funded and shielded by member nations. To take an example that should have given even the most internationalist-oriented
American government contemplating entrusting its military and diplomatic decisions to the UN, Syria was elected to the Security Council a
month after the September 11 attacks, assuming a two-year term in the
seat reserved for Arab countries.[1]

Ambivalence toward American unilateralism and multilateralism can
also be seen in the responses of NATO and the European Union after

[1] Serge Schmemann, "Syria Is Likely to Join U.N. Security Council," *New York Times*,
October 7, 2001; Ginger Thompson, "Mexico Wins U.N. Security Council Seat, Strengthening Fox's World Role," *New York Times*, October 9, 2001.

September 11. The NATO allies were quick to evoke section 5 of their charter, confirming that the attack on the United States was an attack on all nations in the alliance. The support for a military response was wide and deep, and France and Germany were ready to provide more than logistical support in the intervention in Afghanistan. However, the dynamics within this multilateral organization quickly took on more varied shadings. The United States spurned the involvement of French or German forces in what was the first sign of the administration's aversion to involving it transatlantic partners in decision making. Meanwhile, Tony Blair gave himself the role of America's most forceful and active advocate, somewhat distancing Great Britain from its European partners in NATO and the European Union. Even as concern was voiced in Europe that Britain was advancing its "special relation" to the United States at the expense of its European ties, Blair's diplomatic successes gave Britain an undeniably more important role on the international scene and in the eventual intervention in Afghanistan. As the Americans and Britons mobilized, the Europeans themselves were in the midst of an assessment of the relative weakness of their own military strength. Various plans put forward by France, Britain, and Germany regarding Afghanistan received lukewarm response from other members. *Le Monde* editorialist Daniel Vernet succinctly, and ironically, summarized European indecisiveness: "There hadn't been enough prior discussion; the members had the feeling that the crucial things were happening elsewhere, on the ground and in the air: Europe as such was not involved in the military operations and, even if it wanted to be, it didn't have the means."[2]

The ultimate shape of the offensive against al Qaeda and the Taliban was an ad hoc mixture of unilateral, multilateral, and international action. The United Nations legitimized the American-orchestrated offensive; NATO provided crucial diplomatic and logistical support; Pakistan was enlisted as a crucial ally; various Middle Eastern countries provided logistical support or delicate acquiescence; Russia, quite deftly and in its crudest national interests, became a partner against the Taliban; and so on. Meanwhile, the armed forces were American and British, the Americans were in command, and the Afghan Northern Alliance did almost

---

[2]Daniel Vernet, "L'éclipse de l'Europe dans la guerre," *Le Monde,* October 20, 2001.

all the dirty work. To repeat, the war against terrorism took the shape of the United States acting in self-defense *and* in the interests of many other nations. That in effect *is* the international consensus. It has none of the imagined neatness of a community of nations deliberating together and all the messy risks of the world's superpower defending itself as it will.

Terrorism as Symptom

The other two antiwar arguments—the one laying the blame for the attacks on the United States itself, the other on the irresolution of the Palestinian-Israeli conflict—hold that removing the ostensible causes of terrorism takes precedence over a war on terrorism. It is argued that causes rather than effects must be treated. Such arguments claim a more thoughtful, more analytic awareness of history and try to look at the present with a retrospective gaze. But such a perspective can paradoxically drain history of its temporality. The Islamic terrorist networks erupted onto the stage of contemporary history revealing an unforeseen capacity for destruction. They effectively created something new, altering the chains of cause and effect. Fredric Jameson's strange-sounding equivocation a few weeks after the attacks—"I have been reluctant to comment on the recent 'events' because the event in question, as history, is incomplete and one can even say that it has not yet fully happened"— has the merit of admitting that a historical explanation of causes and effects in the present is fraught with uncertainties and opacities. However, since political decision making, unlike historical reflection, takes place within just such an arena of relative opacity and inescapable uncertainty, his stance simply constrained him to forgo *political* judgment altogether. The September 11 attack was an event, and it put the imperative of debilitating the terrorist networks—their command centers, cadres, training programs, and financial support—ahead of addressing past causes. The dangers of the terrorist offensive were immediate, while the task of removing its causes or sources is the work of decades. The United States bears much responsibility for creating this predicament, and the Bush administration proved recklessly inept at fashioning a longer-term policy. Nevertheless, their primary responsibility was to respond with the means at their disposal to the newly posed threat to the political and economic well-being of Western and Middle Eastern nations.

Placing responsibility or blame for the September 11 attacks on American policy itself—something that has been expressed, whether as overt argument or underlying sentiment, by many on the left—obscures what is *now* at stake in international affairs. Principally it fails to understand al Qaeda and kindred groups, however much the United States was involved in their creation in its anti-Soviet strategy in the 1970s and '80s in Afghanistan. Consider the frequently voiced idea that the terrorists are motivated by poverty, resentment toward the West, or humiliations inflicted by Western powers. Poverty has historically given rise to many different forms of political organization (or apolitical submission); resentment and humiliation are in themselves not a cause of political action. On the contrary, resentment and humiliation are politically cultivated sentiments, and they are cultivated as part of an organized strategy for the mobilization of the masses. Al Qaeda is not the fruit of Arab poverty, Israel's stifling of Palestinian national aspirations, or the United States's selective support of some repressive Arab regimes, like Saudi Arabia, and assault on others, like Iraq. Al Qaeda is an outgrowth of Islamic fundamentalism, and it has organized its will to destruction and its mobilization of the masses in a bid for power whose ultimate aims, insofar as they have a concrete shape, are to establish theocratic states that would deprive the masses of secular education, civil liberties, and political and religious freedoms and subject women to harsh, state-organized oppression. These aims do not arise from grievances and humiliations; rather, the grievances and humiliations are cultivated to legitimate the aims.

The politics of historic grievance has an insidious appeal in contemporary politics. Serbs found motivation for ethnic cleansing, the devastation of cities, genocide, and mass rape out of a six-hundred-year-old grievance over the Turkish slaughter of Serbian knights and Prince Lazar in the battle of Kossovo Polje. Jewish fundamentalists expand their settlements on the West Bank motivated by six thousand years of suffering and a fervent desire to hurry the coming of the messiah. Osama bin Laden voices humiliation in the name of Muslims and Arabs for the defeat of the Ottoman Empire, the establishment of the state of Israel, the Arab countries' defeat in their wars with Israel in 1967 and 1973,

Egypt's peace settlement with Israel, and the presence of Western economic interests and military forces on the Arabic Peninsula. These cultivated humiliations and grievances should not be confused with meaningful injustices. Who would argue, besides Serb nationalists, that Turks are responsible for the Serbs' destruction of Sarajevo or the Srebenica massacre? Who would argue, except for the Jewish extremists themselves, that Palestinian aspirations or the intifada is the true cause of the Israeli settlers' apocalyptic intentions in their occupation of the West Bank? The argument, or the feeling, that the September 11 attack was an expression of Arab humiliation or grievances against the American role in the world comes down to the same twisted reasoning.

One of the baldest statements of this view came from Saskia Sassen:

> We may think that the debt and growing poverty in the south have nothing to do with the violence in New York and Washington. But they do.
>
> The attacks are a language of last resort: the oppressed and persecuted have used many languages to reach us so far, but we seem unable to translate the meaning. So a few have taken personal responsibility to speak in a language that needs no translation.[3]

For groups like al Qaeda, terrorism is not a language of last resort; it is their action of first resort, a primary strategy for organizing the allegiance of masses who lack effective avenues of political organization and protest within their own countries. Terrorism is not an "expression" of anything. Sassen's own choice of expression was bad metaphor and a mystification. While her suggestion that "a few have taken responsibility to speak" by flying jets filled with passengers into the Twin Towers was perhaps provoked, like Susan Sontag's paean to the terrorists' courage, by President Bush's statements on their cowardice, it remains empty and confused. I find it impossible—indeed, absurd—to speculate on whether a suicidal mission requires courage, but a moral and political reflection on self-sacrifice, especially murderous self-sacrifice, has to ask *what* the perpetrator is taking responsibility for. Certainly not global poverty or international debt! The symbolism in which Islamic suicide bombers envelop their act turns, from all we know, on their belief that

[3]Saskia Sassen ["A Message from the Global South"] *Guardian,* September 12, 2001.

they are at once raining deadly vengeance on infidels and propelling themselves into a virgin-filled paradise. Courageous or not, Mohamed Atta's act had nothing to do with taking responsibility. It needs only be contrasted with those suicidal acts that have been an expression of protest and conscience designed to evoke a sense of horror in those who witnessed them, like the self-immolation of Buddhist monks in South Vietnam in the 1960s. Sassen misses altogether the relevant question of responsibility when it comes to suicide attacks like those in New York, Washington, and Israel, namely, the responsibility of those who, whether in the hills of Afghanistan or the back streets of Ramallah, recruit and prepare young men for martyrdom. Theirs is a political and moral responsibility that not only deserves no validation but reveals how a religiously fueled politics of ultimate ends transforms politics itself into naked manipulation and murder.

Sassen's remarks raise the troubling question of what leads an otherwise insightful and sensitive intellectual to make such bankrupt statements. The answer, I fear, lies in an attitude that has become a part of the moral-political common sense among many segments of the Left. The attitude arises not from a want of morality but an excess of moral certainty, namely, the belief that the West (or the North in Sassen's vocabulary) is so essentially an agent of oppression, impoverishment, and exploitation that virtually any political movement or act seemingly directed against it acquires some aura of validity. Through this moral lens, geopolitical conflicts seem a grand melodrama: the exploitations of the West versus the suffering of the Other. What drop from political scrutiny are the exact nature, shape, and purposes of the organizations purported to be agents of resistance, opposition, or liberation. The simple fact that they exist is taken to be an indictment of the West, the North, capitalism, or globalization. Their ends and means, their motivations and organization, their actual impact on those they mobilize, end up mattering less than that they are in the symbolism of the melodrama anti-West, anti-North, etc.

Chomskian Certitudes

Another kind of moral certainty animates Noam Chomsky's *9–11*, the collection of interviews he gave and revised in the month following the attacks. Chomsky has no illusions about the aims of terrorists and dismisses any plausible connection to globalization: "As for the bin Laden network, they have as little concern for globalization and cultural hegemony as they do for the poor and oppressed people of the Middle East who they have been severely harming for years."[4] He sees the terrorist attack itself not as a direct but an indirect consequence of American policy, stressing in particular, on the basis of polls and reporting in the Middle East, that the United States's embargo and air strikes against Iraq and its support of Israel stir Arab antagonism and enhance sympathy for terrorist organizations.

When it comes to voicing his opposition to a military response to the terrorist attack, he argues from a broad historical and moral perspective on American foreign policy in the last half century. Chomsky's positions have a tenor of uprightness and rigor. This quality is no doubt what gives his arguments their persuasiveness for those who not only share his criticism of American policy but also admire the idealism of his political judgment. I share many of his criticisms of American policy but reject the idealism.

The criticisms bear on the U.S. record of thwarting the self-determination and sovereignty of many countries, often by undermining democratic forces and orchestrating violence against the political opponents of repressive regimes, even against whole populations. The list is shamefully long: Guatemala (1954), Iran under the Shah, Vietnam, Argentina, Uruguay, Chile, the Philippines, Indonesia, Nicaragua and El Salvador, South Africa, and more. The provocative questions of conscience Chomsky poses have a compelling variant and an absurd one. The compelling question: Don't we see that our own country, now so terribly wounded by a terrorist attack, has itself consistently been an agent and supporter of terrorism in other countries? The absurd one: How can the United States justify a military response to an attack on *its* population and sovereignty when it itself has visited unjustifiable violence on so many other nations, directly and indirectly through its coordinated support of counterinsurgents, dictators, and death squads?

Chomsky calls for a simple adherence to one of the United States's own definitions of terrorism: "the calculated use of violence or threat of

[4]Noam Chomsky, *9–11* (New York: Seven Stories Press, 2001), p. 31.

violence to attain goals that are political, religious, or ideological in nature. This is done through intimidation, coercion, or instilling fear." His own position rests on consistency and literalness: "Everyone condemns terrorism, but we have to ask what they mean. . . . I use the term in the literal sense, and hence condemn all terrorist actions, not only those that are called 'terrorist' for propagandistic reasons."

But does this moral stance yield a justifiable political one?

To argue this question, let us look at how Chomsky uses his criticism of American actions in Nicaragua during the Reagan years to make his case against a war on terrorism after September 11. I choose this reference point because I agree with his critical denunciations of American involvement in Central America in the 1980s. It was unjustifiable from the beginning and devastated the Nicaraguan people and their political institutions; the United States gave logistical and material support to a war against civilians and a legitimate government, all as part of Reagan's pursuit of the Cold War.

In fact, this reminder of American policy in Nicaragua can be amplified beyond what Chomsky discusses in his interviews. The fact that many officials in the Bush administration played a role in Reagan's Nicaragua policy should make us all the more wary and vigilant when it comes to assessing their conduct of the war on terrorism. President Bush issued an executive order on November 1, 2001, significantly curtailing access to presidential papers beginning with the Reagan administration; it is little more than a ploy to shield those members of the current administration who participated in the Reagan-Bush administrations between 1980 and 1992 from any unwanted examination of their judgments and actions in Central America, Iran, and the Gulf War.[5] Unlike the aftermath of the Vietnam War, neither the policymakers nor the public have ever doubted the aims and outcomes of the aggression in Nicaragua, and Bush's executive order will undoubtedly postpone any retrospective debate for years.

Chomsky's argument, though, does not go in this direction of sharpening our historical awareness in order to scrutinize the nation's leaders and actions in the present crisis. Instead, he disqualifies any action

[5]See Richard Reeves, "Writing History to Executive Order," *New York Times,* November 16, 2001.

in the present in light of the past. He does so through a somewhat circuitous argument that is heavy with irony. I reconstruct the argument as follows:

Chomsky rejects in principle the use of military force in self-defense and, beyond that, disputes that the September 11 attacks were an act of war. He illustrates the principle by asking, What if Nicaragua and others had responded to the United States on the same principle that the United States uses to justify a war against al Qaeda and the Taliban in Afghanistan?

> When countries are attacked they try to defend themselves, if they can. According to the doctrine proposed, Nicaragua, South Vietnam, Cuba, and numerous others should have been setting off bombs in Washington and other U.S. cities, Palestinians should be applauded for bombings in Tel Aviv, and on and on. It is because such doctrines brought Europe to virtual self-annihilation after hundreds of years of savagery that the nations of the world forged a different compact after World War II, establishing—at least formally—that the resort to force is barred except in the case of self-defense against armed attack until the Security Council acts to protect international peace and security. Specifically, retaliation is barred.

As to the facts of the situation since September 11, Chomsky dismisses out of hand that the terrorist attacks put the United States "under armed attack": "Since the U.S. is not under armed attack, in the sense of Article 51 of the UN Charter, these considerations are irrelevant." (He does not consider the obvious question of whether those attacks are a sign of a willingness and capacity to carry out further attacks.) Hence, from the standpoint of "the fundamental principles of international law," any unilateral or multilateral military offensive is without justification. The September 11 attacks were an act of terrorism subject to international law. How then should the United States proceed? In answer to a question about his claim that the United States is itself a "leading terrorist state," Chomsky replies, "The most obvious example, though far from the most extreme case, is Nicaragua," and with withering irony he uses Nicaragua's response to American actions in the 1980s as a model for how the United States should respond to September 11:

Nicaragua in the 1980s was subjected to violent assault by the U.S. Tens of thousands of people died. . . . The effects on the country are much more severe than the tragedies in New York the other day. They [i.e., the Nicaraguans] went to the World Court, which ruled in their favor, ordering the U.S. to desist and pay substantial reparations. The U.S. dismissed the court judgment with contempt, responding with an immediate escalation of the attack. So Nicaragua then went to the Security Council, which considered a resolution calling on states to observe international law. The U.S. alone vetoed it. They went to the General Assembly, where they got a similar resolution that passed with the United States and Israel opposed two years in a row (joined once by El Salvador). That's the way a state should proceed. If Nicaragua had been powerful enough, it could have set up another criminal court. Those are the measures the U.S. could pursue, and nobody's going to block it.

Such would seem to be the core of Chomsky's argument: the United States should pursue the mechanisms of the United Nations and international legal instruments. But his various comments and criticism do not in fact precisely add up to that. "There is surely virtually unanimous sentiment, which all of us share, for apprehending and punishing the perpetrators, if they can be found." *If they can be found*—that of course was the question of the hour: not only whether they could be found, but how. There could be, Chomsky argues, "an international effort, under the rubric of the UN, to apprehend and try [bin Laden] and his collaborators. It's not impossible that this could be done through diplomatic means, as the Taliban have been indicating in various ways." Chomsky's assessment of the Taliban was at best utterly naïve, since it was quite clear by mid-October that the Taliban, who were being propped up by al Qaeda as much as they were giving it sanctuary, considered, or feigned to consider, the possibility of turning against bin Laden only because the United States was threatening to topple them from power. Chomsky underscores his trust in the Taliban by approvingly quoting the Indian writer Arundhati Roy: "'The Taliban's response to U.S. demands for the extradition of bin Laden has been uncharacteristically reasonable: produce the evidence, then we'll hand him over. President Bush's response is that the demand is non-negotiable.'" I do not in fact think

Chomsky is naïve; his ventriloquizing of the Taliban's desperate posturing is merely a way station in his ultimate argument that there is a lack of evidence against bin Laden: "the documentation is surprisingly thin. Only a small fraction of it bears on the Sept. 11 crimes, and that little would surely not be taken seriously if presented as a charge against Western state criminals or their clients." Thus, the argument comes full circle: the United States does not have a right to self-defense in responding to September 11 because it is not under armed attack; the United States, which was a terrorist and criminal itself in Nicaragua, should, like Nicaragua, pursue justice through international courts and the Untied Nations; the apprehension of bin Laden should proceed through diplomatic channels with the Taliban; there is insufficient evidence to persuade the Taliban or an international court to extradite and convict him.

Chomsky's brief for inaction is thus complete, rather like that of public defenders who have done their job if the judge says, "Case dismissed." What is left? A principled statement, filled with ironic denunciations, that international politics should be ruled by international courts and that a nation's right to self-defense is an uncivilized archaism that is, moreover, as regards the United States, rank hypocrisy. There is no *justified* course of action for the United States to take against al Qaeda as things stand; there is no justifiable course for the *United States* to take against al Qaeda in any case.

Chomsky's political stance on September 11 is, in short, vacuous. Behind it lies a consistent and coherent intellectual and moral framework. What needs therefore to be questioned is this very consistency and coherence.

He argues from the standpoint of an international structure of legal and moral norms in which every nation is held accountable to the same evaluation of the rightness of its actions, the same definition of terrorism and criminality, and the same sanctions for violating the sovereignty of other nations and supporting or perpetrating acts of terrorism as defined. However, this legal-moral structure does not exist. Chomsky does not merely argue that it *should*; he argues from a perspective *as though it did*. It is that gesture that I find unacceptably idealistic in matters of historical analysis and political judgment.

The gap between ideal and reality has to be taken into account to understand the new international conflict that emerged on September

11, and to judge what the United States and other nations should do about it. The ideal Chomsky adheres to—one in which all nations are subject to the same constraints by international organizations and courts before which they are all equals—is far from being shared by the nations of the world. Such international legal, moral, and political institutions exist today only partially and in the midst of an international "order"—that is, really, a field of conflicting, uneven forces—in which nation-states and multinational alliances exist alongside subnational movements (including "ethnic" groups vying for political inclusion or dominance within an established state or for separation from it) and supranational movements (including the international terrorists networks). When Chomsky, for example, dismisses the idea that the September 11 attacks put the United States under armed attack as defined by the UN charter, he ignores precisely the novelty of terrorism within this international reality: al Qaeda is an armed force that does not need to be a state in order to undertake a war against a state. That it was flourishing under the protection of a state, Taliban-ruled Afghanistan, was in turn what justified the United States in treating the Taliban as a belligerent in the wake of September 11.

Unlike the ideal of an international legal-moral order, the existing dynamics of international politics and power cast the United States in an ambiguous role. As I have already suggested, the international community, including the United Nations, has in effect established a consensus that the United States should lead the diplomatic and military offensive against terrorism. American unilateralism is intermixed with international and multilateral decisions. (Chomsky himself has acknowledged this fact, albeit negatively, in his opposition to the Gulf War, the bombing of Serbia, and the NATO action in Kosovo.) The ambiguity of the United States's superpower standing after the Cold War stems from the contradictions within and between two roles:

First, as a nation-state more powerful than any other, the United States acts unilaterally in pursuit of its national interests and self-defense; those pursuits have long been distorted by contradictions in America's double commitment to democracy and to the expansion of capitalism.

Second, as the strongest nation within an international order mediated by the United Nations and alliances like NATO, the United States is repeatedly turned to lead the multilateral resolution of international

conflicts, whether by diplomacy or force; since the international institutions devoted to making binding diplomatic and legal decisions are
merely partial, and since the global imperatives of the international
economy are not able to be controlled by any nation, including the
United States, this second role introduces new contradictions (and
hypocrisies) not just among unilateralism, multilateralism, and internationalism, and between democratic commitment and capitalist expansion, but also in the meaning of sovereignty, human rights, civil war,
and the rule of law within and between nations.

Hardt and Negri's Empire

The resulting ambivalence in U.S. foreign policy is addressed in intriguing terms by Michael Hardt and Antonio Negri in *Empire*.[6] In their
vocabulary, the American role in the world is an admixture of *imperialism* inherited from modern European nation-states and the newer
mantle of the leader of *empire,* that is, globalization understood as a
new economic, social, and political order. They exemplify their distinction between imperialism and empire with two historical watersheds. Whereas the Tet offensive in the Vietnam War in 1968 marked
"the irreversible military defeat of U.S. imperialist adventures" (179),
the Gulf War in 1991 "presented the United States as the only power
able to manage international justice" (180). In the new era the globalized form of capitalism combines with an incipiently post–nation-state
order mediated through the United Nations and international law to
give the United States "the responsibility of exercising an international
police power . . . *not as a function of its own national interests but in the
name of global right.*"

Although the difference between the Vietnam War and the Gulf War
might seem nothing more than the difference between the Cold War
and post–Cold War forms of American hegemony—the position that
underlies Chomsky's analyses—Hardt and Negri complicate this pic-
ture. They insist on the difference between the Cold War and post–
Cold War models of American hegemony, and they then relate this difference to a longstanding contradiction in American history between a

[6]Michael Hardt and Antonio Negri, *Empire* (Cambridge: Harvard University Press, 2000).

republican form of empire and modern nation-state imperialism. It is a question of two distinct visions of the relation between conquest and governing. Republican empire expands territorially in order to open itself to new members and so to new social forces. Nation-state imperialism controls lands and peoples in order to exploit them.

In the broad outlines of American history, this distinction corresponds to the difference between the western expansion led by Jefferson and Jackson and the imperialist enterprise that was inaugurated by the Spanish-American War in 1898. In the latter, America became a nation-state imperialist ruling over foreign peoples; in the former, it sought to expand its territory and open itself to economic and political changes. Hardt and Negri evoke the work of J.G.A. Pocock to underscore the Machiavellian moment in American republicanism. It is their own reading of Machiavelli that yields the concepts by which they then understand the nature of republican empire in America's past:

> First of all, there is the Machiavellian concept of power as a *constituent power*—that is, as a product of an internal and immanent social dynamic. For Machiavelli, power is always republican; it is always the product of the life of the multitude and constitutes its fabric of expression. The free city of Renaissance humanism is the utopia that anchors this revolutionary principle. The second Machiavellian principle at work here is that the social base of this democratic sovereignty is always conflictual. Power is organized through the emergence and the interplay of counterpowers. The city is thus a constituent power that is formed through plural social conflicts articulated in continuous constitutional processes." (162)

Noting that Machiavelli and the republican revolutionaries in America were alike intrigued by imperial Rome, Hardt and Negri draw from Machiavelli a political principle for republican empire: "Machiavelli defined as expansive those republics whose democratic foundations led to both the continuous production of conflicts and the appropriation of new territories" (166). The western expansion of America across the continent in the nineteenth century becomes their model of the "notion of sovereignty" inherent in empire: the "tendency toward an open, expansive project operating on an unbounded terrain" (165).

This principle of democratic sovereignty and this aspiration to empire are utterly distinct from modern nation-state sovereignty and imperialism:

> The fundamental difference is that the expansiveness of the immanent concept of sovereignty is inclusive, not exclusive. In other words, when it expands, this new sovereignty does not annex or destroy the other powers it faces but on the contrary opens itself to them, including them in the network. What opens is the basis of consensus, and thus, through the constitutive networks of powers and counterpowers, the entire social body is continually reformed. . . . Empire can only be conceived as a universal republic, a network of powers and counterpowers structured in a boundless and inclusive architecture. (166)

This conceptual scheme does the heavy-lifting in Hardt and Negri's project: it is meant to account for the phases of American history, explain America's role in foreign affairs and global capitalism today, and anchor a new vision of revolution.

American history has the shape of the inauguration, loss, and recovery of the ambition to empire: in the beginning, the project of territorial expansion to spread democracy across the continent; then, at least from the time of the Spanish-American War, a regression to the European model of nation-state imperialism until this project collapsed with the war in Vietnam; and finally, after 1989, leadership of the new, internationally sanctioned project of empire.

American foreign policy today vacillates because the United States still clings to some obsolete assumptions of nation-state sovereignty and imperialism even as it takes on the new, or revamped, mission of republican empire. As capitalism developed into its global form, it required a new form of political rule and a new policing power to flourish. It has fallen to the United States to provide this rule and power, Hardt and Negri suggest, not primarily because the end of the Cold War left the country with military superiority but because the conception of a political and economic order that can extend itself without frontiers or borders is the deep-seated, though residual, American experience of empire since the time of Jefferson: "Empire can only be conceived as a universal republic, a network of powers and counterpowers structured in a boundless and inclusive architecture." Within American history, then, the original project of continental expansion has returned, thanks to the global developments of capitalism, in the form of American leadership of postmodern empire. Conversely, capitalist globalization has found, thanks to the republican experience of expansion in American history, its needed postmodern political form.

Finally, the conception of empire as a "universal republic" and "network of powers and counterpowers" is the utopian red thread of history. Hardt and Negri tug at this thread, pulling it loose from the fabric of American constitutional and diplomatic history to reweave it into the prophetic tapestry of their own utopian vision: counterglobalization as the revolution *in* Empire. They want to reclaim the republican idea of empire from, and set it against, its current form as the empire of global capitalism. They envision a global antiglobalization. For, if republican empire is necessarily universal, then the "counterpowers" to which it gives rise in its "network of powers and counterpowers" signal the "immanence" of the possibility of revolutionary change. The old models of revolution are obsolete. Since the United States is not embarked on nation-state imperialism, national liberation movements are no longer a viable model. Since no nation-state, including the United States, controls the development of capitalism within its own borders, proletarian revolution in the nineteenth-century sense is irrelevant. Models of local resistance to globalization are doomed for the same reasons.

Hardt and Negri's perspective on America's global role represents a stark contrast to Chomsky's. They see the United States answering the transnational imperatives of capitalism and the call of the international community to extend and police the new economic-political order, whereas Chomsky sees the United States as the ultimate rogue state, setting itself against international justice and the sovereignty of other nations. He sees U.S. action as a violation of the moral and legal principles that other nations cherish, whereas they see the United States as obeying the amoral laws of contemporary history. Chomsky's politics would rein in U.S. global adventurism and subject U.S. actions to international sanctions and tribunals. Hardt and Negri's politics welcomes American-led globalization and its republican-universal tendencies as the seedbed of new possibilities of revolution and community.

The Multitude and Prophecy

The common sense of the American left today stands on two legs: Chomskian moralism and a diffuse form of the Marxist legacy. The fact that Chomskian moralism and Marxist theory are incompatible, intellectually and politically, does not keep them from freely combining in the sensibility of the Left. Political sensibility does not require consistency. It is a set of perspectives, prejudices, and social perceptions that enables the forming of opinions on the events of the day, and given the complexity and unpredictability of events it may be that a political sensibility is all the more responsive to events, all the more flexible and intuitive, the less consistent it is.

The persistence of the Marxist legacy is so diffuse today that it is mostly manifest simply in the vague assumption that genuine social justice would require, somehow, the overthrow of capitalism. This assumption perennially renews the belief—the hope, the expectation—that some particular revolt or rebellion is the harbinger of systemic change. Perhaps the Chiapas uprising, perhaps strikes in France or riots in Los Angeles, or the intifada—any of these can reawaken from a distance the feeling that capitalism is not permanent, that there are movements in the world that make universal social justice imaginable, even fleetingly.

There occasionally arises a powerful intellectual effort to refurbish Marxist theory and discern anew within capitalism the outlines of its possible overthrow. Hardt and Negri's *Empire* is such an endeavor. They cleave to the Marxist tradition in their twofold aim: they aspire to create a theoretical synthesis that links economy, politics, and culture into the unity of a system, and they soothsay the prospect of revolution in various current revolts and rebellions, deciphering in them the immanence, if not imminence, of systemic change. The Zapatista and Palestinian movements, the L.A. riots, and the 1995 general strike in France are in fact Hardt and Negri's list of recent revolts in which they see the sort of "constituent struggles, creating new public spaces and new forms of community," that prefigure revolutionary change. The supposed commonality is that each is at once purely local and yet reacts to some global feature of capitalism. Their lack of any actual connection—they do not "communicate" with one another—gives the new theory its task: "Recognizing a common enemy and inventing a common language of struggles are certainly important political tasks, and we will advance them as far as we can in this book, but our intuition tells us that this line of analysis finally fails to grasp the real potential presented by these new struggles" (57). The difficulty of relating these movements becomes the sign of their real relation: "Perhaps the incommunicability of struggles, the lack of well-structured, communicating tunnels, is in fact a strength rather than a weakness—a strength because all of the movements are immediately subversive in themselves and do not wait on any sort of external aid or extension to guarantee their effectiveness" (58). Are these movements the prefiguration, the hope, of revolution, or are they merely figures, metaphors, of the theorists' hope for revolution? Hardt and Negri's language, as in the passage just quoted, dwells in the zone of ambiguity where that question need not be answered.

The intellectual pattern of *Empire,* along with its rhetorical design, sheds light on the political sensibility of the Left today not because Hardt and Negri give voice to a deep revolutionary yearning in oppressed and excluded peoples (which is how they understand their project), nor because the existing Left of activists and intellectuals would embrace their theory (which they would not uniformly do), but rather because they develop a theory of contemporary capitalism that elaborates, often poetically, the Left's usually submerged, doubt-ridden hope that somehow the desperate, often bloody struggles of indigenous peasants, ghetto residents, or stateless peoples offer a *figure* of global social transformation.

The rhetoric, or poetic, of *Empire* is inseparable from its ideas. Hardt and Negri have chosen the genre of their discourse. It is the manifesto. The Communist Manifesto is the model of course, but they also reach back to Spinoza: "Today a manifesto, a political discourse, should aspire to fulfill a Spinozist prophetic function, the function of an immanent desire that organizes the multitude" (66). *Empire* truly is a manifesto in the Marxian tradition, despite its length. Perhaps not since Guy Debord's *Society of the Spectacle* has this genre been revived so self-consciously and supply. As with Debord, their writing imitates Marx's in rhythm and phrasing. And its intellectual scheme, like Marx's, has all the coiled equipoise of a tae kwon do master ready to defeat any errant idea or fact that comes its way. For anyone steeped in the Marxist tradition, *Empire* is a marvel of familiarity. For those who are not, it must seem either an exhilarating invitation to comprehend the whole of today's baffling global social and political reality—or else mere speculative bombast. I approach it from the first of these three attitudes. *Empire* is a marvel and all too familiar.

Its theses are by turns prophetic, hortatory, totalizing, attuned to the mass, voluntarist, and hopeful beyond hope:

> *Prophetic:* "The spectacle of imperial order is not an ironclad world, but actually opens up the real possibility of its overturning and new potentials for revolution" (324).
>
> *Hortatory:* "What we need is to create a new social body, which is a project that goes well beyond refusal. Our line of flight, our exodus must be constituent and create a real alternative. Beyond the simple refusal, or as part of that refusal, we need also to construct a new mode of life and above all a new community. This project leads not toward the naked life of *homo tantum* but toward *homohomo,* humanity squared, enriched by the collective intelligence and love of the community" (204).
>
> *Totalizing:* "In the same way today, given that the limits and unresolvable problems of the new imperial right are fixed, theory and practice can go beyond them, finding once again an ontological basis of antagonism—within Empire, but also against and beyond Empire, at the same level of totality" (21).
>
> *Attuned to the masses:* "Today a manifesto, a political discourse, should aspire to fulfill a Spinozist prophetic function, the function of an immanent desire that organizes the multitude. There is not finally here any determinism or utopia: this is rather the radical counterpower, ontologically grounded not on any 'vide pour le futur' but on the actual activity of the multitude, its creation, production, and power—a materialist teleology" (66).
>
> *Voluntarist:* "Being republican today, then, means first of all struggling within and constructing against Empire, on its hybrid, modulating terrains. And here we should add, against all moralisms and all positions of resentment and nostalgia, that this new imperial terrain provides greater possibilities for creation and liberation. The multitude, in its will to be-against and its desire for liberation, must push through Empire to come out the other side" (218).

Hopeful beyond hope: "We have to recognize where in the transnational networks of production, the circuits of the world market, and the global structures of capitalist rule there is the potential for rupture and the motor for a future that is not simply doomed to repeat the past cycles of capitalism" (239).

Now, this last evocation of a future that might escape "the past cycles of capitalism" hides the question that anyone in Hardt and Negri's intellectual tradition has to ask: What does the survival of capitalism, with its protean transformations and ever more globalized innovations, reveal about the past cycles of *Marxism?* For Marxism has resurrected itself many times before in the wake of events that confounded its expectations and its theory. Each time the theory resurrected the aspiration to theoretical totalization and the prophetic discovery of a new agent of revolutionary change: the Leninist party (Lukács), the peasant masses (Mao), the national liberation of colonized peoples (Fanon), the unified revolt of workers and students (Debord), the countercultural revolt of youth (Marcuse), the fragmented principle of hope itself (Bloch).

Prophetic discovery is central to all the theses-tropes I quoted above—each of which, by the way, is the concluding statement of a chapter. The manifesto itself claims to function as "an immanent desire that *organizes* the multitude." In remarkable contradiction to the self-assurance of the prophecy is the vagueness of the multitude's immanent desire. Even more, the revolution that promises to be republican and universal has no vision of the multitude's self-government: "we will still not be in a position—not even at the end of this book—to point to any already existing and concrete elaboration of a political alternative to Empire" (206). But even the inability of their theory to conceive what the multitude, whose immanent desire the same theory knows with such certitude, might actually create, what new forms of government might emerge, even that failure does not in the eyes of the authors refute or even cast doubt on the theory. Rather, it confirms their conviction. Why? Because just such a gap has its authoritative precedent in Marx! "At a certain point in his thinking Marx needed the Paris commune to make the leap and conceive communism in concrete terms as an effective alternative to capitalist society. Some such experiment or series of experiments advanced through the genius of collective practice will certainly be necessary today to take that next concrete step and create a new social body beyond Empire" (206). As they await the revelatory event, they do not investigate the actual fate of the world's various upheavals and revolts. That the people of Chiapas might find a new social contract with modern Mexico, that without a new federal commitment to urban policy African Americans in Los Angeles will remain mired in poverty and crime, that a negotiated settlement of the Israeli-Palestinian conflict might open the way for Palestinians to determine their destiny—such undoubtedly reformist questions do not interest Hardt and Negri. They might well argue that such solutions are rendered impossible by the dynamics of global capitalism itself. But what is more evident about their position is that such preoccupations would deprive their own revolutionary imagination of the Indians of Chiapas, urban blacks, and Palestinians as *images* and *figures* of revolutionary hope.

It is the enjoyment of such images and figures of the revolutionary "multitude" that the Left must learn to resist, just as it must learn to forgo moral absolutism. Chomskian moralism and revolutionary poetics *do* meet in the political sensibility, offering hope and rectitude in place of political judgment.

Political words and notions and acts are not intelligible
save in the context of the issues that divide the men
who use them.   —ISAIAH BERLIN

IRAQ: DELIRIUM OF WAR,
DELUSIONS OF PEACE

The Idealism of Means

Only one outcome of the invasion of Iraq met with nearly universal approval: the fall of Saddam Hussein. Yet "regime change" had been the
most contested, confused, ill-defined, and poorly justified reason for the
war. This paradox underscores how disoriented American and European
foreign policy was (and remains) when it comes to understanding the
responsibility and power of Western democracies in the face of tyrannical regimes on the international scene. Confusion afflicted not only the
Atlantic alliance in the buildup to the war in Iraq but also the United
Nations. And it afflicted thinkers and writers.

It is worth reflecting on the arguments of intellectuals and nations
regarding the war in Iraq because few, if any, of the questions posed by
the prewar debates have been resolved. No reliable precedent has
emerged to square principles and practices. I am going to frame the discussion by juxtaposing Jürgen Habermas and Paul Berman. Their opposing views of the war in Iraq are highlighted by the fact that they both
belong to an antitotalitarian, left-liberal intellectual heritage. Both are
committed to political liberalism and social democracy; both support a
global expansion of human rights and maintain a sharply critical view
of the Nixon-Kissinger tradition of American foreign policy "realism";
both supported "humanitarian intervention" in the Balkans. When it
came to Iraq, they disagreed over the invasion *and* over the principles of
intervention.

Habermas found that the end did not justify the means in the overthrow of Saddam Hussein. It is not exactly clear, however, in what sense he meant this. Is it that good ends *never* justify compromised, ambiguous, harmful means? If so, he would have to retreat to a position he has always eschewed, namely, an idealism that asserts moral truths no matter how askew from real-world problems and choices. Or is it that the advantages of ending *this* dictator's rule did not outweigh the *actual* damage done to the progress of international and cosmopolitan law when the United States and Britain acted without UN approval? If so, Habermas would have needed to suggest how the various judgments are to be made: What was the extent of Hussein's tyranny? How was international law likely to have progressed *without* the intervention? How was the one consideration to be weighed against the other? He left all these questions unaddressed. This uncharacteristic lack of clarity underscores the uncertainty the Iraq crisis provoked for political thought and moral reflection. Habermas's own way of expressing the difficulty deserves scrutiny. Writing just a few weeks after American troops entered Baghdad, he recalls the televised image of Saddam Hussein's statue being pulled down. It evoked an "ambivalence" in the "moral sentiments" of the European spectator whose revulsion at the "shock and awe" of an "illegal war" mixed with the gratification of this image of a people's liberation from the "terror and oppression" of a regime whose "criminal nature" was beyond doubt. *Illegal war* versus *criminal regime.* How to choose? Habermas warns against letting moral feelings overwhelm political judgment: "An illegal war remains an act contrary to international law, even if it entails desirable consequences from a normative standpoint." At the same time he acknowleges a hard question. Since "catastrophic consequences can in effect delegitimate a laudable intention, why can't laudable consequences acquire a capacity for legitimation aposteriori?" Where, in short, does the viable normative standpoint lie?[1]

The United States justified its action as taking over a role the United Nations has largely failed to fulfill. Even if the aims of such "a self-conferred hegemonic power—a 'hegemon'—"were realizable, "a hegemonic unilateralism nonetheless entails secondary effects that . . . are

---

[1]Jürgen Habermas, "La statue et les révolutionnaires," *Le Monde,* May 2, 2003.

undesirable normatively" with regard to its own criterion of "bettering the world by virtue of liberal ideas." Habermas rightly points to the abuses of prisoners' rights at Guantánamo as evidence that Bush's unilateralist vision of his "war presidency" jeopardizes liberal values themselves. But that still does not answer the hard question, *illegal war* or *criminal regime?* Instead of responding, Habermas caricatures the moral feelings arosed by the American intervention: "Saddam overturned from his pedastal remains the sole argument, the symbol of the new liberal order over the whole region." The relevant question—*did putting an end to Saddam Hussein's dictatorship justify aposteriori an intervention in breach of international law and UN decision-making procedures?*—concerns the fall of *Saddam Hussein,* not the fall of Saddam Hussein's *statue.*

Habermas simply sticks to the principle that the established UN rules of deliberation, decision, and action are self-evidently more valid than outcomes in international relations. He does not finally engage the opposing position: If a dangerous tyrant can be overthrown with a reasonable expectation that he will be replaced by a stable and more liberal, more democratic regime, international law should permit it; and in the absence of the appropriate international law, responsible democratic nations should initiate such an undertaking anyway. Those are normative claims that can be extrapolated from the NATO intervention in the Balkans. Habermas tacitly rejects them by referring back to his own support of the intervention in Kosovo. Despite the lack there too of Security Council authorization, "the hope that a legitimation could be recovered aposteriori—as it indeed was—was able to be based on three factors": namely, military action put a stop to ethnic cleansing, promulgated the universal injunction to aid those in need, and was carried out by a coalition of democratic countries. What began to become apparent already in 1999 according to Habermas was a divergence in the normative principles of the "continental European powers" and the "Anglo-Saxon powers": the Europeans "strove to draw the lesson of the Srebrenica tragedy by reducing the gap between effectiveness and legitimacy," while Britain and America "maintained as a satisfying normative objective the act of extending, even outside our borders and by force if necessary, the Western liberal order."

In Habermas's view the underlying antagonism between these two perspectives, which he will characterize, respectively, as "Kantian cosmopolitanism" and "liberal nationalism," broke out anew in the controversy over Iraq. He implies that now a choice must be made between these contrary sets of values, and it is in effect self-evident to him that toppling a foreign tyrant to enable a liberal political order is a less defensible norm than upholding existing international law and procedures.

The Idealism of Ends

The journalist Paul Berman forcefully argued the opposing view. For him, the principle of liberal interventionism to overthrow tyranny is throughly justified, since it expands on the precedent of humanitarian intervention. The expansion from genocide and ethnic cleansing to tyranny is, moreover, in keeping with the bitter lesson of the twentieth century on the dangers of appeasing or underestimating the destruction and injustice wrought by totalitarianism. He devotes a large part of *Terror and Liberalism* to arguing that Baathism and Islamism are "two branches of a single impulse, which was Muslim totalitarianism—the Muslim variation on the European idea."[2]

Berman also sharply criticizes Bush's diplomacy and arguments on behalf of the war. Why, he asks on the eve of the war, did Bush turn to the spurious reason of Saddam Hussein's ties to al Qaeda and the hardly urgent reason of weapons of mass destruction? The aim of deposing a totalitarian tyrant and creating the possibility of a liberal Iraq was justification enough. It appealed to a universal principle and, in extending the right or duty of "humanitarian intervention" established in the Balkans crisis, addressed values widely shared in Europe and elswhere. Berman astutely points out that as Bush repudiated international law and even the Geneva Conventions, he squandered his persuasiveness: "Laws, formal treaties, the customs of civilized nations, the legitimacy of international institutions—these were the dross of the past, and Bush was plunging into the future. And, as he plunged, he had no idea, nobody in his administration seemed to have any idea, that international law, human rights, 'Europe,' and humanitarianism had will-nilly become the

---

[2]Paul Berman, *Terror and Liberalism* (New York: W. W. Norton, 2003), p. 60.

language of liberal democracy around the world."[3] By not making that case for a liberal intervention to allies and the international community, and especially to Muslims, Bush undercut the prospect of success. The same error had already compromised the war in Afghanistan. "The part about going after bin Laden and Al Qaeda needed no elaboration. . . . But there was another aspect of the Afghan War—the part about overthrowing totalitarianism and bringing the benefits of a free society. Bush spoke and even, as I say, acted on these principles. But the hesitations and the cautions, the miserly husbanding of resources, the obvious reluctance to do more than the minimum—these things undid his every remark."[4]

There is much that I admire in Berman's thought and much I agree with in his argument. Saddam Hussein's regime in Iraq was a dangerous totalitarian state whose overthrow democratic countries had an obligation to pursue. The question was when and how. The invasion was in my view an error first and foremost because it deflected from the far more pressing war against Islamic terrorism that had been successfully *inaugurated* in Afghanistan only to be subordinated to the preoccupation with Iraq and virtually abandoned on account of the administration's utter indifferrence to nation building. The case for the urgency of an Iraq intervention was never made; it was simply manufactured by the American troop buildup and the most spurious of the administration's arguments. And *how* should an intervention be pursued? On this question Berman's argument is marred by a blindspot. Bush's bad reasons were more than a drawback of the intervention. For when a democracy takes up arms, the failure to articulate the reasons, whether through inability or, worse yet, refusal, destroys the validity of the war itself. The failure to justify the war left it unjustified. Beyond the fact that the bad arguments alienated allies and world opinion, they committed the American people to a course of action whose motives and purposes they would ultimately not be able to sustain.

Much of the passion in Berman's view is fueled by his belief, also quite valid, that leftist intellectuals in the twentieth century accommodated

[3]Ibid., p. 193.
[4]Ibid., p. 196.

and apologized for communist totalitarianism out of a naïve refusal to believe in its horrors and a confused idea that criticizing capitalist society entailed defending communist regimes. He makes a powerful case against doing anything similar with Islamism and Baathism. And he was right to see confusion and appeasement in the attitudes of many protesters against the war in Iraq: "Some of the protesters invoked "just war" theory. In "just war" theory, to invade a country in order to stop massacres currently underway is deemed perfectly just. But to send in armies to rescue the survivors after the massacres have ended is deemed unjust. The marchers in 2003 gazed at Iraq and saw plainly that the massacres had come to an end. (And logically so: The survivors had been clubbed into submission, and no Iraqi was going to rise in rebellion ever again.) And the marchers therefore swelled in indignation."[5]

There is no lack of conviction in Berman's support of the war, but the passionate intensity derived, I think, from another, more troublesome source as well. His writing evinces a surplus of excitement. There is too much pleasure in the prospect of American miltiary supremacy being used for a just cause. It resembles the excess of enthusiasm in the rhetoric of the neoconservatives. My explanation is based on an intuition, but I think it is sound. The war in Vietnam weighs in the background of both Berman's and the neoconservatives' enthusiasm. Anyone who held antiwar and patriotic sentiments simultaneously during the Vietnam War—a combination common among the antitotaltarian, antisectarian Left—lived the United States's misguided adventure in Southeast Asia, the atrocities and hypocrises, the lies and the disregard for the lives of civilians, as a painful rent in the national conscience, and so in one's own conscience. It remained unhealed through Nixon and Kissinger's complicity in the overthrow of democracy in Chile and the disaster of Cambodia and on through Reagan's orchestration of the decimation of El Salvador and Nicaragua, all in the name of anticommunism and containment. What a relief, what a lifting of the burden, if now, after Vietnam, Laos, and Cambodia, the United States actually used its military might for something unambiguously good and right, liberating and democratic! Let America be America again! Let it be rep-

---

[5]Paul Berman, "Silence and Cruelty: Five Lessons from a Bad Year," *New Republic,* June 28, 2004.

resented once again by GIs landing on the beaches of Normandy rather than grunts burning hootches in the Mekong Delta! As for the neoconservative architects of the war in Iraq, they have undoubtedly carried their own burden of guilt over the last few decades. After all, they were complicit in the phony liberation of Grenada, the antidemocratic brutalities in Nicaragua and El Salvador, and of course the Iran-contra scandal—not to mention Pinochet, Marcos, and South African apartheid. Hypocrisy and blood must weigh on the minds of such principled intellectuals as William Kristol and Robert Kagan, not to mention Reagan-era government servants like the perennial Paul Wolfowitz and the rehabilitated Elliot Abrams. They too expressed the idea of overthrowing Saddam Hussein with a surplus of satisfaction, an excess of enjoyment. What Berman and the neoconservatives shared was this risky indulgence in a guilt-assuaging enthusiasm for arms in the service of right.

So, what is wrong with such enthusiasm? Where is the fault in assuaging guilt after years of bearing it in your soul? What better relief than seeing the nation's terrible instruments of destruction turned to a just cause instead of compromised ones? The problem is quite simply the one that returns me again and again to Max Weber's "Politics as Vocation." There is no redemption in violence. Violence may be necessary, it may liberate, it may serve good ends. Therein lie some of its possible justifications. It is never redemptive. Indeed, it is never simply *just,* for it is always pregnant with tragedy. The sense of tragic possibility was not only absent from George W. Bush's awareness going into the war in Iraq, it also faded from the consciousness of those intellectuals who indulged in the idealistic, romantic, revolutionary belief that the absolute supremacy of U.S. miltary strength could only do good, could only do the right thing, if turned to the task of ending tyranny. In terms of Weber's classic distinction, these intellectuals succumbed to the belief that the rightness of the *ultimate ends* would spare us *responsibility* for evil.

Habermas denounces the intervention for lacking legal justification; Berman supplies liberal ideals to provide the missing justification. If Habermas is right that Bush's determined unilateralism undermined international law, it does not make Berman wrong that the overthrow of a tyrant for the sake of liberalism is a worthy ideal to guide international action. Berman risks letting the ultimate *end* of a just cause compromise the ethic of responsibility; Habermas risks compromising the ethic of responsibility by idealizing the *means* of international procedures at expense of liberating a people from despotism. Had Habermas, the proceduralist of Kantian cosmopolitanism, advocated European support of the intervention he would have advanced the cause of antitotalitarianism; had Berman, the idealist of liberal intervention, opposed a unilateral American intervention he would have advanced the cause of cosmopolitanism. This cat's cradle of ideals and justifications, procedures and outcomes, reflects something of the complexity of questions of war and peace in an age where unprecedented threats from terrorism and totalitarian ideologies and regimes strain existing international arangements.

Neither Left nor Right

Perhaps Berman and Habermas represent too sharp a contrast—the idealism of ends *versus* the idealism of means. Would a dose of pragmatic values clarify things? That is, a consideration of what is necessary, achievable, and prudent. If so, it will also refocus the value that both the idealism of ends and the idealism of means try to bracket out: national interests. A reasoned case for the war in Iraq, carefully argued and thoroughly documented, did in fact get made before the invasion, though not by the Bush administration. Its author was Kenneth M. Pollack, who spent several years as a CIA military analyst of Iraq and Iran and then served on the National Security Council in 1995–1996 and 1999–2001 during the Clinton administration. His 2002 book *The Threatening Storm* was frequently cited by nervous editorialists who were prepared to support the war but were frustrated, if not outright frightened, by the poor case Bush was making to the nation and the world. Why couldn't the president and his administration present the case for war as forcefully and cogently as Pollack?

Pollack's line of argument is an essential benchmark because it hews to the values that are likely to guide any American administration facing the question of using military force. He privileges neither international law nor the ideals of liberal intervention. Unlike Habermas, Pollack does not accept that a president of the United States has a higher duty to existing international law than to the nation's security and interests. Unlike Berman, he does not accept that the United States has a duty to liberate oppressed people from tyranny. He argued within a realist perspective grounded in national interest: Saddam Hussein's regime, because of its development of weapons of mass destruction and potential effect on oil supplies, posed a sufficiently serious threat to American security that it was in our national interest to overthrow him. Pollack's realism is neither cynical nor devoid of higher values, and he holds an expansive interpretation of American national interest. He argued that United States would enhance its own security by committing itself to a democratic reconstruction of Iraq: "rebuilding Iraq politically and economically holds out the greatest likelihood for stability in Iraq and the region. . . . If the United States is going to employ major military force, it should 'seek to create an end-state that is significantly more positive than what existed before hand."[6]

Pollack follows the established protocols for justifying military action: "It is the inadequacy of all the other options toward Iraq that leads us to the last resort of a full-scale invasion."[7] He made the case by demonstrating in a richly detailed account of the history of Saddam Hussein's regime and American dealings with it that every previous and existing American policy toward Iraq, especially from the end of the Gulf War up until 2002, had failed to loosen his hold on power or slow his development of weapons of mass destruction. To let the status quo unfold could lead to a far worse predicament than the perils of an invasion: "Relying on pure deterrence to keep Saddam at bay once he has acquired a nuclear arsenal is terrifyingly dangerous. It is likely to find us confronting a Hobson's choice of either allowing Saddam to make himself the hegemon of the Gulf region (and in effect or actuality controlling global oil supplies) or else fighting a war with him that could escalate to nuclear warfare. . . . Unfortunately the only prudent and realistic course of action left to the United States is to mount a full-scale invasion of Iraq to smash the Iraqi armed forces, depose Saddam's regime, and rid the country of weapons of mass destruction."[8] That Saddam Hussein aspired to be "the hegemon of the Gulf region" and had undertaken a concerted effort to acquire nuclear weapons was beyond doubt.

---

[6]Kenneth M. Pollack, *The Threatening Storm: The Case for Invading Iraq* (New York: Random House, 2002), p. 397.

[7]Ibid., p. 335.

[8]Ibid.

No one—including the United States and France, Israel and Syria, Germany and Russia—doubted the threat between September 2002 and March 2003 as the international debate over military action unfolded. The debate in the United Nations in January and February centered on whether the newly resumed weapons inspections should continue—and whether Saddam Hussein's threat to peace was imminent. When it became apparent almost immediately after the March invasion that Saddam Hussein had no viable nuclear program, indeed had no viable program of biological or chemical weapons or the missiles capable of delivering them, the carefully argued rationale crumbled. There was not an *imminent,* even an eventual, Iraqi threat. The invasion proved neither *prudent* nor *realistic*—nor *the only* course of action left to the United States.

What Saddam Hussein was hiding was that he had nothing to hide. Pollack, writing in the *Atlantic Monthly* (January 2004) after David Kay reported on the Iraq Survey Group's vain search for weapons, identifies the two most plausible explanations for why Hussein kept up this subterfuge despite the economic sanctions and then the threat of war. Possibly, as Iraq's former UN ambassador Tariq Aziz reportedly told his American captors, Hussein wanted to maintain the illusion in the Arab world that he possessed the capacity to dominate the Middle East and attack Israel. Even more plausibly, in Pollack's view, he thought his grip on Iraqi society depended on the illusion of holding weapons of mass destruction. His was a tyrant's madness, and a tyrant's weakness. The show of power disguised a fear of losing power.

The implacable logic of "the case for invading Iraq," as Pollack subtitled his book, was not really lost on the Bush administration. They followed it quite closely. However poorly they made the case themselves, they apparently considered and rejected the more "idealistic" justifications. According to Bob Woodward's reconstruction of the preparation for the war, the option of going to the United Nations with a proposal to overthrow Saddam Hussein because he was a tyrant was considered but rejected on the assumption that the Security Council would reject it out of hand: "It was clear that few other countries would support such an effort. The terrorism case seemed weak or unprovable, and the issue of seeking regime change because Saddam was a dictator or a particularly brutal despot would not get to first base. It would be both loudly and quietly laughed out of the United Nations, which had its share of countries with one-man rule. WMD was really the only one that had any 'legs,' Rice said, because at least a dozen resolutions on Iraq's WMD already had been passed and to one extent or another ignored by Saddam Hussein."[9]

While the administration may well have been following a rationale parallel to Pollack's, they lost sight of the sort of realism that tempered his analysis and especially his prognosis. It is as though Bush concluded that *since* everything else had failed, an invasion would succeed. Something of the same non sequitur pervades Pollack's book itself. For it is the relentless piling up of past failures that lends an air of reasonableness to the idea of undertaking the most extreme and risky action yet: invasion, toppling a government, occupying a foreign land. Moreover, where Pollack understood what sort of commitment of troops, resources, and time might be required of an American occupation, the Bush administration gave almost no thought to the postwar at all.

There were arguments *against* the war that also turned on pragmatic questions of necessity, feasibility, and prudence. I have already referred to a major one, namely, that the focus on Iraq deflected from the war on terrorism, especially the hunt for al Qaeda, and the stabilizing of Afghanistan. To this could be added the neglected priority of pushing the Palestinians and Israelis toward a resolution of their conflict. Add also the incoherence of focusing on Iraq while the true seedbeds of Islamic terrorism and weapons proliferation—Saudi Arabia and Pakistan—were two of our favored allies!

Two other pragmatic arguments against the war were important; the one turned out to be quite valid and the other surprisingly invalid. The valid argument was the majority position on the Security Council, articulated most forcefully by France, namely, that it was better to pursue the UN weapons inspections as a way of determining the actual extent of Saddam Hussein's threat while neutralizing him and perhaps even loosening his grip on power. The fact that he had no hidden weapons of mass destruction gives great weight in retrospect to this argument.[10]

---

[9] Bob Woodward, *Plan of Attack* (New York: Simon & Schuster, 2004), p. 220.

[10] So strong is Pollack's antipathy toward France and the Security Council that even as he admits his own shock at the mistakes in prewar intelligence, that is, the revelation of the extraordinary imprudence of the assumptions behind the war, he still cannot admit that these

Positive results would not of course have been automatic. The pressure of imminent military action was required to make Saddam Hussein continue his half-hearted but sufficient cooperation with the UN inspectors, and how much time to grant the Iraqi regime would have been difficult to negotiate among the allies and in the United Nations, especially since the U.S. military preparation in the region was massive (and therefore very expensive) and geared to attack in the spring before the hot weather.

The other antiwar argument was that the economic sanctions were an effective brake on Saddam Hussein's power. This view turned out to be no more accurate than the prewar intelligence on weapons of mass destructions. The sanctions had been, as human rights critics had long claimed, devastating for the well-being of Iraqis. However, contrary to the architects of the sanctions and the antiwar voices claiming their effectiveness, the whole program *strengthened* Saddam Hussein's hold on power. David Rieff, reporting in the *New York Times Magazine* in June 2003, detailed how this happened. Especially during the so-called food-for-oil years, the contracts permitted with foreign companies included lucrative markups and systematic kickbacks. Favored nations saw their corporations profit; Saddam Hussein controlled the kickbacks. (The complicity of UN officials in the corruption was eventually revealed.) The rationing system created the illusion of Hussein's largesse to the people and gave the tyrant a new method of surveillance and control over an already fearful, cowed, and impoverished population: "Every Iraqi head of household had to have . . . a ration book, issued by the Ministry of Trade, which named every immediate family member and listed the precise quantities of foodstuffs to which the bearer was entitled. Every food agent had a computerized list from the Ministry of Trade of the persons he was supposed to supply with these staples. What this meant in practice was that the regime could maintain a database on

revelations confirm that the Security Council and France were right in February and March. This dogmatic blindspot in his otherwise clear and analytic vision leads him in fact to draw the opposite conclusion: "The shameful performance of the United Nations Security Council members (particularly France and Germany) in 2002–2003 was final proof that containment would not have worked much longer; Saddam would have reconstituted his WMD programs, although further in the future than we thought." Kenneth M. Pollack, "Mourning After: My Debate with Bill Galston," *New Republic,* June 28, 2004.

every citizen and update it, without recourse to the security services or even a network of paid informants. It was a secret policeman's dream—and it was all provided, however inadvertantly, by the sanctions the United States and Britain had conceived as a way of limiting Saddam Hussein's power."[11]

What is the lesson of this episode that enhanced Saddam Hussein's power even as it impoverished Iraqis, to the point of raising infant mortality? It suggests that sanctions are doomed to failure in an oil-rich, import-dependent economy under the tyrannical control of a modern miltary and administrative apparatus. Yet, it is on tyrannies that the international community is most likely to impose economic sanctions. Rieff points out that South Africa under apartheid is often held up as an example of the success of sanctions but is perhaps simply the exception that proves the rule. There is a further, more pointed conclusion to draw. South Africa had a market economy, and though the state was de jure racist and held the black majority in thrall, it was not a dictatorship. The openness of the political system, however limited, and of the economy allowed the sanctions to work. Sanctions against dictatorships without an open economy, like Iraq or Cuba, merely tighten state control. A further, devastating effect—which is, I think, a part of the still untold story of the failed U.S. occupation in Iraq—is that under such circumstances sanctions destroy the hold that the middle class and working class have in civil society. Since those are the classes that foment and sustain democratic aspirations, sanctions undercut the very possibility of democratic revolution by devastating and demoralizing the middle and working classes.

In light of the pragmatic and the principled arguments, the debate over intervention in Iraq did not yield a cogent political or ideological divide. Habermas and Berman diverge over the war in a clash between the idealism of means and the idealism of ends, but their two stances do not separate as left and right. So, too, pragmatists or realists, who resist such idealism in either form, divided over whether the war was necessary, prudent, and in the national interest. While the discourse the Bush administration used to justify its military-diplomatic deeds can be traced to a handful of neoconservative tracts, the crucial events themselves—

[11]David Rieff, "Were Sanctions Right?" *New York Times Magazine*, July 27, 2003.

the war in Afghanistan, the ongoing war against international terrorism, and the overthrow of Saddam Hussein and occupation of Iraq—have engendered controversies that do not neatly fit to the traditional political antagonisms of the Left and the Right. Since the end of the Cold War and the 1991 Gulf War, the divisions of political opinion on foreign policy have generally not followed any coherent right-left pattern.

There is, rather, a patchwork of ideas and position-taking that confirms the Arendtian premise that political judgment is never simply a matter of applying a principle to a situation, perhaps especially when questions of war and democracy, sovereignty and international law, are at stake. Among many prominent European leftists, for example, the political thought of the Nazi legal scholar Carl Schmitt has become a touchstone for understanding sovereignty as the power to declare the "state of exception" that suspends the rule of law; when not preoccupied with vague utopian ideas of a society beyond the aporia of the legal foundation of lawlessness and the lawless foundation of legality, this strand of contemporary political thought broadly condemns American foreign policy as exemplifying the ultimate (or inevitable) identity of sovereignty and the "state of exception," sovereign power and lawlessness. At the same time, other European leftists defended the principle of national sovereignty to vehemently oppose the NATO intervention in the Balkans and the bombing of Belgrade. Still others, like Habermas and the French socialist politician and founder of Doctors Without Borders Bernard Kouchner, supported "humanitarian intervention" as a justification for limiting or overriding national sovereignty in order to prevent and punish ethnic cleansing and genocide; with the war in Iraq, however, Kouchner, like Berman, used the same line of reasoning to support the overthrow of Saddam Hussein, and Habermas to oppose it. The handful of French intellectuals on the liberal left who joined Kouchner in supporting the war, especially André Glucksmann and Pascal Bruckner, agonized over how to give the Bush administration's actions a principled articulation as a defense of freedom and human rights. Old intellectual allies in the European struggle against totalitarianism, like Daniel Cohn-Bendit and Adam Michnik, divided over the justification for the war. To Cohn-Bendit's vehement opposition to Bush, the Polish editor and former dissident simply replied, "A bad administration, bad reasons, a very good intervention!"

The Atlantic Misalliance

Nations themselves did not follow a clear political or ideological blue-print for deciding war and peace. The war was incubated and then orchestrated by the neoconservative hawks in the administration of a Republican president, but the leading U.S. ally was Britain's New Labor prime minister. The war was strenuously opposed by France's center-right president and foreign minister and Germany's social-democratic chancellor and green foreign minister, all of whom were in turn embarrassed by the prowar stance of the new democracies of Eastern Europe.

If there was no left and right in the dispute over Iraq, was it then the war camp versus the peace camp? That dichotomy is even less enlightening, since it begs the critical question: What do war and peace mean to Western democracies today? On this question public opinion in the West also plunged into the fog. One of the most remarkable aspects of this volatile period was the near-unanimity of public opinion on each side of the Atlantic. George W. Bush enjoyed overwhelming support for the war, while European public opinion was almost unanimous in its opposition. That American and European opinion bifurcated was less telling than the absence of conflict within each. For the unanimity in each case was sustained by powerful misconceptions: as American and British troops launched their offensive, some 45 percent of Americans believed that Saddam Hussein was the architect of the September 11 attacks on the World Trade Center and the Pentagon, and a quarter of the French public sided with Iraq against America. Delirium of war on one side of the Atlantic, delusions of peace on the other.

Politicians seldom dare to dispel the misconceptions of their own supporters, so it is unsurprising, though no less disturbing, that Bush went to war quite content to keep Americans in the dark about the difference between al Qaeda terrorism and Baathist tyranny, while Jacques Chirac, enjoying the backing of 80 percent of European opinion as the champion of the so-called peace camp, did nothing to challenge the ignorance of those who streamed into the streets celebrating Saddam Hussein and likening Bush and Ariel Sharon to Hitler. If the misinformed views had been marginal, such corruptions of public debate would not matter much, but delirium and delusion were the glue of unanimity for war and for peace.

How then did the Western democracies decide for war or peace?

The United States, France, and Great Britain are the longest-standing and most powerful democracies in the West, and Germany's economic and democratic achievements since World War II, and since reunification, have made it the cornerstone of European stability and prosperity. These four democracies are also allies of the first order. Yet their diplomatic showdown over Iraq was hardly a deliberation over the means to realize a shared end or over competing ends within mutually understood means; it did not even have the shape of knights at their roundtable, with the United States playing the first among equals. Nor did the leaders' positions derive from any debate within their respective polities that was commensurate with the grave and complex issues at hand:

George W. Bush, whose questionable electoral victory in 2000 rendered his presidency relatively aimless until September 11, 2001, willfully pushed the issue of Iraq to the pitch of crisis and manipulated American public opinion by casting war in Iraq as a part of the war on terrorism.

Tony Blair set a course that flew in the face of British public opinion and seemed to delight in casting himself in the role of the statesman moving his country against the current of the majority.

Gerhard Schroeder risked diminishing Germany's international role by appealing to Germans' instinctive pacifism in an election campaign he was otherwise in danger of losing.

Jacques Chirac, only a few months before the diplomatic battle in the United Nations, had seen his dismal showing of 22 percent in the first round of France's presidential elections turn into a landslide of more than 80 percent in the second round, where the vote was a referendum against the extreme Right rather than an endorsement of his own presidency. His 80 percent support lacked substance until the Iraq crisis, where it took the simplistic form of No War, and if anything the French president let his public support bewitch his judgment at the crucial junctures of his dispute with the United States and Britain.

And so the four greatest Western democracies wended their way along four crooked paths to a war that left them divided and an occupation that would have had a far greater chance of success if they were united.

When France entered into negotiations with the United States over Resolution 1441 in October and November 2002, it shouldered two tasks: securing a UN-directed disarmament of Iraq and discouraging American unilateralism. Quite consistent in themselves, the two tasks soon became the horns of a dilemma, for the simple reason that the Bush administration did not come to the United Nations with the goal of disarming Iraq peacefully. Bush was trying to balance the antagonistic viewpoints within his administration, most obviously represented by the differences between Secretary of Defense Donald Rumsfeld and Secretary of State Colin Powell. Powell thought a war against Iraq should have the legitimation of UN approval, following the precedent laid down by the president's father in the Gulf War and consistent with the diplomatic route followed in preparation of the invasion of Afghanistan and overthrow of the Taliban; Rumsfeld thought the UN was unnecessary, indeed a hindrance to American interests. The balance Bush struck was to go to the UN for authorization and meanwhile affirm that the United States would go to war in Iraq with or without UN approval. UN legitimation was a desirable cover for U.S. policy rather than an indispensable framework. No one in the administration, including Powell, looked to the inspections to disarm Iraq; the inspections were simply to create an internationally recognized justification for military intervention.

France was acutely aware of the stakes. Chirac had warned Bush for several months that France opposed preventive war and unilateral American intervention (two tenets of the Bush doctrine), considered a settlement of the Israeli-Palestinian conflict the highest priority in the Middle East, and worried that a preoccupation with Iraq and neglect of Afghanistan would weaken the war on terrorism. The neoconservative hawks took the opposite view on all these issues. Once Bush decided to take the question of Iraq to the UN, France behaved as though it were negotiating with Colin Powell on a shared premise that the United Nations was the rightful framework for dealing with Iraq and that disarmament via inspections, backed by the continuing economic sanctions and the new threat of force, was preferable to war. Powell was negotiating with the French as though the shared assumption was that as soon as Saddam Hussein inevitably failed to provide proof of complete disarmament the war would begin. For France, military force was a last resort; for the United States, it was inevitable and imminent and simply awaited the UN's certification of Saddam Hussein's noncooperation.

The mismatch of American and French assumptions was tacitly acknowledged by being woven into the ambiguous language of Resolution 1441: the French gave up their original demand, spelled out by Chirac in a *New York Times* interview on September 9, 2002, for two resolutions (the first giving Iraq three weeks to admit UN inspectors "without restrictions or preconditions," and a second, if Saddam Hussein did not cooperate, to decide if military force was called for), and the Americans gave up their demand for an automatic authorization of force and accepted further meetings of the Security Council to assess Iraqi compliance with the inspections.

Once Hans Blix began reporting mixed results to the Security Council—yes, Iraq was cooperating with the inspectors; no, it was not providing proof of having disarmed—the debate between the United States and France shifted and hardened. The United States reiterated its position that Saddam Hussein's history of weapons development, his expulsion of UN inspectors in 1998, and the underestimation of his weapons by previous international inspections meant that now he had to be held to the letter of his agreements with the United Nations and provide clear evidence that he had no biological, chemical, or nuclear weapons or ongoing research. Was Saddam Hussein proving that he had disarmed? Blix was saying, Not yet. France posed a different question: Was there any reliable evidence that Saddam Hussein had sufficient weapons of mass destruction to threaten the security of his neighbors and the West? Blix was saying, Not yet. For France, the absence of a clear threat eviscerated the justification for war; moreover, as long as the inspectors were at work in Iraq Sadddam Hussein was effectively neutralized.

The dispute between France and the United States erupted on January 20, 2003, Martin Luther King, Jr., Day, in the highly publicized rift between Colin Powell and France's foreign minister Dominique de Villepin. It was high diplomatic drama. Powell felt sandbagged when he answered Villepin's call for a special meeting of foreign ministers in the Security Council to discuss the war on terrorism only to be confronted with a debate on Iraq by the German foreign minister Joschka Fischer and others. Afterwards, Villepin announced to the press that France saw nothing to justify war in Iraq and would contemplate using its veto on the Security Council to block it. Powell's widely reported rage and feeling of betrayal were a clear sign that he had counted on France ultimately to facilitate UN approval of military action. Villepin's gesture now exposed Powell to the machinations of his rivals in the Defense Department, since Rumsfeld and the neoconservative hawks were hardly displeased at Powell's failure to win UN support. It freed their hand from UN mandates and weakened Powell's position within the administration at a time when Rumsfeld was upstaging him by continually making pronouncements on foreign affairs, not just defense policy, and had gone a long way in having the Defense Intelligence Agency outmaneuver and eclipse the State Department's intelligence bureau. The hawks could also now scapegoat France as a way of loosening the public's instinctive belief in the importance of the United Nations.

In a well-documented report on this episode, gleaned from several high-level anonymous sources, *Le Monde* (March 27, 2003) surmised that Chirac and Villepin chose this moment to confront the United States because of three developments: first, a French diplomat's meeting with Condoleeza Rice and Paul Wolfowitz convinced them that the Bush administration was planning military action no matter what; second, they were alarmed by the rapid deployment of U.S. troops in preparation for an invasion; and, third, they anticipated that Hans Blix's report the next week would weaken their own position, presumably because it was going to show Saddam Hussein's cat-and-mouse game with the inspectors and certainly not provide proof of disarmament. France abruptly stopped assuming, or dropped the pretense, that negotiations with Powell were based on shared premises.

While Bush's talk of "regime change" and his threat to act with or without the United Nations had been a part of the landscape all along, it seems most plausible that the French preview of Blix's report made them fear that their effort to make the inspections process work was in jeopardy. As it turned out, this anxiety was not well-founded. Blix's report of January 27 stirred more doubt than conviction about the immediate justification for war; Powell's theatrical presentation on February 4 of proof of Saddam Hussein's weapons of mass destruction was utterly unpersuasive except in the United States; and Villepin's own arguments in the debate of the Security Council on February 14 gave so much credence to continuing the inspections that even the Bush administration had to take a pause in its criticisms of the UN and the Old Europe. France reached the height of its influence on the international debate at this moment. Ironically, Villepin's threatened veto was not the source of this success. On the contrary, the UN debates showed that his dramatic gesture had been unnecessary. Worse, it undercut the French position from then on. And, worse yet, Jacques Chirac went on to align France's position with Germany's. Gerhard Schroeder had taken Germany out of the Iraq debate altogether the preceding summer when he sounded the chord of German pacifism during his reelection campaign and declared that Germany would not support military action even if it had UN approval. Nothing could have been further from the position originally taken by France: Iraqi disarmament via UN inspections backed by the credible threat of force.

Between October and February France was the most effective brake on American unilateralism and a forceful advocate of international agreements and action. It had deftly exploited its seat as a permanent member of the Security Council to create a tentative equilibrium with the United States in the eyes of the international community and world opinion; it gave a kind of international moral and political counterweight to the obvious superior strength of the United States. What was lost by brandishing the veto (prematurely and publicly) and aligning with Germany (needlessly and self-contradictorily) was the possibility of putting a brake on Washington while *simultaneously* increasing pressure on Iraq. It certainly could be argued that such a feat was doomed from the beginning, since France and the other members of the Security Council lacked a pledge from the United States to be bound by Security Council decisions. But if that had been France's assessment (rather than simply a reasonable fear), it would not have acted as it did from October until January. Once it embraced the decision in November to disarm Iraq, France was obligated not to relinquish that aim even as it strenuously opposed U.S. plans. Contrary to France's intention, the veto threat and alignment with Germany may well have weakened rather than defended the United Nations and its credibility, since the whole Iraq controversy arose in part because Saddam Hussein had for twelve years eluded UN enforcement of its own resolutions.

Diplomatic Intrigues and Political Truths

In the crisis that rattled Europe and the Atlantic alliance in February and March, Tony Blair and Jacques Chirac chose to exacerbate rather than overcome the divisions among European leaders. They exacerbated them by dramatizing them: Chirac aligned with Germany and then Russia against the United States; Blair orchestrated the petition of support for the U.S. position by the Eastern European countries; Chirac chastised the signatories, with special scorn for those countries awaiting admission to the European Union. Meanwhile the Bush administration, emboldened by its success with the American media and public in demonizing France and the United Nations, decided to forgo further UN debate and go to war.

The theatrical split between Blair and Chirac squandered what their views had in common, namely, their opposition to the Bush doctrine, as outlined in *The National Security Strategy of the United States* in September 2002, and more particularly to any unilateral aggression. In fact, France's position since Resolution 1441 that Iraq must disarm via inspections or face military action was not substantially different from Britain's; and Britain's desire for a UN framework was not that different from France's insistence on it. The outlines of a common position between France and Britain (and even Germany, which was not altogether at ease with its self-imposed marginality) were at least imaginable at the time—and look even more compelling in retrospect. Britain and France could have demanded in March that (1) the U.S. agree not to act without UN approval; (2) the intensification of inspections be coupled with a predetermined timetable for intervention; (3) an international plan, not dominated by the United States, be undertaken to reconstruct Iraq in the event of war. Such a British-French position would have expressed what the two governments already had in common: an opposition to the Bush doctrine, a commitment to disarming Iraq via the UN, a desire for maximum international consensus if war was necessary, a stake in post-Hussein Iraq. And it would have addressed what both countries had so far fallen far short of achieving: greater unity in European foreign policy and shared responsibility for change in the Middle East.

Perhaps the most plausible reason that this opportunity was missed is the rivalry between Blair and Chirac, Britain and France, in vying for leadership of the new Europe. For Blair, the advantages of solidarity with France, including the revitalization and forcefulness of the UN and the easing of domestic opposition to his diplomacy, were apparently outweighed by Britain's "special relation" with the United States and his own ambition to lead Europe. For Chirac, the advantages for Europe and the UN of solidarity with Britain apparently did not outweigh the risks of quite likely involving France in a war with Iraq, thus losing the mantle of international hero of peace, and altering the shape of his own ambitions to be the leader of Europe.

As Britain and France failed to discover a new European consensus, and therefore new leverage for steering the United States on a more multilateral, international, rule-governed course of action, the four major Western allies ended up stuck in four irreconcilable stances: American unilateralism, German pacifism, British pragmatic idealism, and French proceduralism. This characterization calls for clarification, since none of the four countries adhered unwaveringly to its principle. Neither prowar nor antiwar postures were altogether as they seemed.

Britain can be credited with the most consistency, if only because its pragmatic idealism is not a principle in quite the same way as the others. Therein lies its great advantage. At the moment of diplomatic crisis Blair moved to the American position, after having consistently pressed on Washington the necessity of UN involvement; he did not so much reverse himself as simply recalibrate the mix of idealism and pragmatism, appealing now to the moral imperative of overthrowing Saddam Hussein even at the expense of the United Nations.

The German and French positions may have appealed to international and cosmopolitan values over against American unilateralism, but both countries were themselves energized by powerful nationalistic trends. The antiwar movement in Germany coincided with a renewed fascination with the "memory" of the devastating Allied bombings of German cities during World War II—actual memories, but more importantly commemorative symbolizations. Antiwar protesters linked Dresden and Baghdad as twin targets of Allied air power; the plausible association of Saddam Hussein with Adolph Hitler strangely metamorphosed into Germans' identification with Iraqis. Germans and Iraqis were alike seen as victims of British-American force. Pacifism may have proclaimed itself in the name of international law, but it found expression—its existential conviction—in a national identity of Innocent Victim.[12] For a nation to seek its identity in, pride itself on, innocence

[12]Andreas Huyssen, *War Burnout: Memories of the Air War* (Barcelona: Centre de Cultura Contemporania de Barcelona, 2004), pp. 339–45. Published in conjunction with the exhibition "At War."

and victimage can seem a self-defeating way to influence the international scene, even a will-to-irrelevance, but the German stance at the same time, in yet another register, was mimicking—whether in mockery, scorn, or envy—Americans' dramatized perception of themselves as victims and innocents since September 11. The mantle of victim has become a symbolic stake in disputes over what is just and unjust. These mimetic-symbolic underpinnings of the German antiwar protests cast doubt on the view of Habermas and others who celebrated the mass protests against the invasion of Iraq as the emergence of a European and global public sphere infused with a commitment to cosmopolitan principles.

The antiwar stance in France was no less complex and contradictory. French diplomacy voiced the absolute priority of international law, cosmopolitan values, and UN authority, but it also clearly chose this route because France holds a veto on the Security Council and can use this privileged position to enhance its national interests in influencing world events and strengthening its role as a leader of Europe. Whatever were the decisive factors that caused France to turn to open opposition of Bush's Iraq policy, its own commitment to proceduralism in the name of international law was *at the same time* a bid for national power and superiority.

Such contradictions and hypocrises in diplomatic strategy do not disqualify the German or French positions as such. They are a reminder, though, that pacifist or proceduralist principles will never fully explain (or determine) German or French diplomacy. Although pacifism mobilized German opposition to the war in Iraq, it did not accurately reflect Schroeder and Fischer's thinking in general. As for Chirac, there is no reason to doubt the sincerity of his devotion to the ideals of international law as the basis of European and Atlantic foreign policy, but his diplomatic maneuvers were designed to strengthen France's position and influence vis-à-vis the United States and other European countries. The committed internationalist was at the same time acting as a shrewd nationalist. As he should! For it seems to me that the lesson to be drawn from German and French diplomacy in the Iraq crisis is that international or cosmopolitan ideals are never in fact altogether separate from national interest, and indeed without that "realist" ballast the international situation would likely be more dangerous than it already is.

Repudiations of the UN Left and Right

The UN's charter and history do not envision an international responsibility to overthrow tyranny. The Soviet Union under Stalin was after all an original permanent member of the Security Council. Recent interventions did not directly confront the question. The United Nations authorized the 1991 Gulf War because Iraq had violated Kuwait's sovereignty. The justifications for the "humanitarian intervention" against Milosovic's Serbia were grounded in stopping ethnic cleansing and were carried out by NATO, not the UN; the reason for avoiding the UN was the certainty of a Russian veto, but even so it is not clear whether such an intervention squares with the UN's fundamental principles regarding national sovereignty. The overthrow of the Taliban was legitimated in the name of the U.S. right to self-defense against a regime that was harboring al Qaeda.

With no clear precedent provided by any of these interventions, the UN negotiations over Iraq at the end of 2002 flirted with a new principle. The consensus that initially formed around Resolution 1441 implied that Saddam Hussein could rightfully be removed from power by military force if such an overthrow were the *consequence* of enforcing the UN resolutions that required him to furnish proof of Iraqi disarmament. The logic was the mirror image of the Bush administration's actual position, which was at bottom crafted on grounds of national security, not international law: Bush argued that Saddam Hussein was a threat to American security (because of alleged ties to al Qaeda and weapons of mass destruction) and more broadly to international security in the Middle East (and therefore to American national interests); he further argued that Saddam Hussein's history of aggression against Iran and Kuwait, massacres of Iraqi citizens, and noncompliance with the United Nations showed the severity of the threat he posed and so justified a preventive war to overthrow him. The conundrum that the Security Council faced (or created for itself) was how to give the American initiative to overthrow a dictator the appearance of an international action to enforce UN disarmament resolutions.

Resolution 1441 has been severely criticized on account of this very duplicity. The criticism, like so many other position-takings on the war in Iraq, has been politically polymorphous, as can be seen in two examples, one neoconservative and the other neo-Marxist.

Michael J. Glennon, writing from a standpoint shared by the neo-conservative hawks in and around the Bush administration, argues that the fracture in the Security Council merely highlighted the moribund nature of the United Nations, whose founding rules on the use of force have long since become empty:

> Since 1945, so many states have used armed force on so many occasions, in flagrant violation of the charter, that the regime can only be said to have collapsed. In framing the charter, the international community failed to anticipate accurately when force would be deemed unacceptable. Nor did it apply sufficient disincentives to instances when it would be so deemed. Given that the UN's is a voluntary system that depends for compliance on state consent, this shortsightedness proved fatal. . . . Massive violation of a treaty by numerous states over a prolonged period can be seen as casting that treaty into desuetude—that is, reducing it to a paper rule that is no longer binding. The violations can also be regarded as subsequent custom that creates new law, supplanting old treaty norms and permitting conduct that was once a violation. . . . If countries had ever truly intended to make the UN's use-of-force rules binding, they would have made the costs of violation greater than the costs of compliance.

As for Resolution 1441 itself, the ambiguous language that allowed the United States to see in it an authorization of force and France to see none "represented a triumph of American diplomacy" according to Glennon and at the same time "a defeat for the international rule of law." Moreover, since Iraq's acceptance of new inspections depended on the American threat of force, the Security Council's supposed "victory of diplomacy" was a "diplomacy backed by the threat of unilateral force in violation of the charter. The unlawful threat of unilateralism enabled the 'legitimate' exercise of multilateralism. The Security Council reaped the benefits of the charter's violation."[13] There is a sobering insight in Glennon's argument: the UN framework not only is at odds with American unilateralism but is also ill-designed to address the most pressing international problems.

Perry Anderson, writing from an ostensibly neo-Marxist standpoint, expresses an equivalent disdain for the United Nations. Rehearsing the

---

[13]Michael J. Glennon, "Why the Security Council Failed," *Foreign Affairs,* May/June 2003.

Bush administration's various arguments for the war in Iraq and the counterarguments of those against it, he concludes that the prevalent antiwar arguments fail to refute the prowar ones because they are based on the assumption that international law and the United Nations are viable counterforces to American unilateralism. False, says Anderson. "No international community exists. The term is a euphemism for American hegemony. It is to the credit of the Administration that some of its officials have abandoned it." Where Glennon considers the Security Council moribund because the UN's charter and rules do not reflect the real basis on which the United States actually makes decisions as the world's dominant power, Anderson considers it bankrupt because it is nothing more than an alibi of American power:

> since the Cold War came to an end, the UN has become essentially a screen for American will. Supposedly dedicated to the cause of international peace, the organisation has waged two major wars since 1945 and prevented none. Its resolutions are mostly exercises in ideological manipulation. Some of its secondary affiliates—UNESCO, UNCTAD and the like—do good work, and the General Assembly does little harm. But there is no prospect of reforming the Security Council. The world would be better off—a more honest and equal arena of states—without it.[14]

Anderson's line of argument is cold comfort to peace movements or governments trying to rein in America under the Bush doctrine. He is so convinced that American hegemony is a fundamental ill and the UN merely its handmaiden that he would apparently chuck not only the Security Council but the Non-Proliferation Treaty as well, considering it "a mockery of any principles of equality or justice—those who possess weapons of mass destruction insisting that everyone except themselves give them up, in the interests of humanity. If any states had a claim to such weapons, it would be small not large ones, since that would counterbalance the overweening power of the latter." He counsels peace movements to think historically and analytically rather than morally and legalistically: "The Iraqi regime is a brutal dictatorship, but until it

---

[14]Perry Anderson, "Casuistries of Peace and War," *London Review of Books* 25, no. 4, March 6, 2003.

attacked an American pawn in the Gulf, it was armed and funded by the West. . . . Arguments about the impending war would do better to focus on the entire prior structure of the special treatment accorded to Iraq by the United Nations, rather than wrangle over the secondary issue of whether to continue strangling the country slowly or to put it out of its misery quickly."

A historical focus, however, hardly leads automatically to a condemnation of the Iraq intervention. For, as Anderson otherwise stresses, American Middle East policy and UN rules and practices were formed in the context of the Cold War. East-West conflict created layer upon layer of hypocritical and contradictory Realpolitik. Western commitments to democracy were always sacrificed in the Middle East for alliances with oil-rich autocracies. With the end of the Cold War have arisen other possibilities. While it is certainly true that the United States would have had no concern with an Iraqi invasion of an oil-poor Kuwait, it is also true that the invasion threatened to destabilize the Middle East and, because of oil, the world economy. Nor was the first Bush's coalition war under UN auspices in itself harmful; if anything, it turned out to set too weak a precedent. At issue in 2003 was not only the question Anderson coyly dismisses as naïve—*Did the actions of the Iraqi dictatorship justify an intervention to overthrow it?*—but also one he does not bother to ask: *Under what conditions is putting an end to a tyrannical regime a valid basis for international or multilateral military action?* For no matter how severely one criticizes the methods, preparation, or prudence of Bush's policy, it included from the beginning the intention of overthrowing a vicious tyrant and establishing a democracy in Iraq. If such a project was not thinkable in American foreign policy until the Cold War ended, all the more reason to celebrate the collapse of Soviet communism.

Where Anderson considers it self-evident that American hegemony is a global harm, Glennon just as dogmatically finds the Bush doctrine to be the self-evident, natural expression of American hegemony. Glennon fails to ask whether alternatives to the Bush doctrine might not give American global leadership a different shape that would more effectively establish alliances, improve national and international security, and foster democracy abroad. He thus leaves one of his own most pregnant statements unexamined and undeveloped: "A new international legal order, if it is to function effectively, must reflect the underlying dynamics of power, culture, and security. If it does not—if its norms are again unrealistic and do not affect the way states actually behave and the real forces to which they respond—the community of nations will again end up with paper rules."

Glennon postpones this task of rebuilding the international legal order into an indefinite future: "Some day policy makers will return to the drawing board." This intellectual dodge imitates in miniature the flaw that is writ large in the Bush doctrine. At a moment in history when the United States found itself the sole superpower, enjoying a liberal democracy that much of the world admires and having succeeded in the course of seven decades in weaving close alliances with the most democratic nations in the world—why would it choose this moment *not* to put forth a new vision of international law? Why in this moment would it actively seek to disrupt the cooperation among these democracies in global affairs, even to the point of estranging its longest-standing allies?

Let's follow Glennon and shine a harsh light on the problems of the post–Cold War international legal order. The 1991 Gulf War adhered to international law but was an anomalous crisis; the interventions in Bosnia and Kosovo cirumvented the UN charter and Security Council procedures; the overthrow of the Taliban stretched international law; and—let us add—the failure in Somalia and abject negligence in Rwanda exposed fatal gaps in international law and tragic indifference on the part of the international community. In Glennon's own terms, then, the interventions since 1991 have begun to rewrite international law via calculated violations and new initiatives; at the same time, the Western democracies' failure to achieve a common framework for the war in Iraq as it had for the intervention in the Balkans and, just as importantly, their passivity in the face of the Rwandan genocide have revealed how incomplete and indistinct the emergent rules (customary or formalized) are. The metaphor of "back to the drawing board" is a mere alibi, since the United States itself, by its actions and inactions, has figured centrally in all these post–Cold War crises, just as it was itself the main architect of the UN charter and the Security Council in the first place. It was, and remains the United States's responsibility to refashion the written and unwritten rules of international relations in collaboration with its allies and the society of nations, rather than spurn laws, treaties, rules, and alliances.

The Hobbesian Nightmare: Occupied Iraq

After all the neoconservative rhetoric against the UN and the American show of unilateralism and "coalitions of the willing," the most serious error in Bush's policy was to attempt an invasion and occupation of Iraq without a very broad coalition if not UN authorization. The invasion could succeed unilaterally, but not the occupation. The occupation showed how ill-equipped the United States is militarily and politically, intellectually and administratively, to rebuild a nation and foster a democracy. The misunderstanding of the task of nation building was rooted in the same assumptions as the will-to-unilateralism: the belief that America's own democracy is legitimacy enough for an intervention designed to bring democracy to an oppressed people. Habermas's worry that America set out to impose democracy imperialistically on others is misplaced. The tragedy of the Iraq intervention lay in the naïveté that makes the administration believe that American democracy makes American force at once effective and benign.

By the time the invasion and occupation of Iraq produces its equivalent of *The Pentagon Papers* it will be startlingly clear that although Iraq in no way resembled Vietnam, the United States made all the same errors it had made forty years earlier. It overestimated the importance of military superiority; it misunderstood the role of nationalism and the dynamics of civil strife; it lacked a coherent and sustainable justification for its involvement; it misled the public on the severity of the threat that the intervention was supposed to remove; it persisted in its initial errors because it could not learn from the experiences and facts that contradicted the original "theory" behind the intervention.

The specific causes and shape of the failed occupation are a physiognomy of intellectual and political negligence. Kenneth M. Pollack, the former CIA military analyst who had made perhaps the best argument for the war in 2002, looked back at the American occupation two years later in amazement. Writing in an issue of the *New Republic* in which a number of figures who had supported the war offered their reassessments and second thoughts in May 2004, a year into the occupation, amidst the Sunni and Shiite insurgencies, Pollack pointedly asks: Why did the administration "dismiss all of the preparations for postwar reconstruction performed by the Department of State, USAID, the intelligence community, the uniformed services, and a host of other agencies, and instead follow Ahmed Chalabi's siren song?"[15] The scope of what the Americans neglected was massive. Why were government buildings, museums, and libraries in Baghdad allowed to be sacked and looted? Why were the coalition forces unprepared and unable to secure civil order? Why were essential public services and utilities left unrepaired? Why were so many mistakes made in failing to reconstitute Iraqi governmental agencies, police, and army?

George Packer, reporting from Baghdad for the *New Yorker*, already assessed the problems of the occupation in great detail in November 2003, foreseeing the crisis that would unfold more and more dramatically through the winter and spring of 2004 and not diminish after the "sovereignty" of a still crippled, strife-ridden Iraq was entrusted in July to the Iraqi Interim Government that had been chosen under tight American supervision. Packer also reported on the origins of the catastrophe. The Defense Department and the vice president's office utterly ignored the research that the State Department had done in 2002 by working with sixteen groups of Iraqi exiles on anticipated postwar problems ranging "from the electricity grid to the justice system": "The Pentagon also spent time developing a postwar scenario, but, because of Rumsfeld's battle with Powell over foreign policy, it didn't coordinate its ideas with the State Department. The planning was directed, in an atmosphere of near-total secrecy, by Douglas J. Feith, the Under-Secretary of defense for policy, and William Luti, his deputy. According to a Defense Department official, Feith's team pointedly excluded Pentagon officials with experience in postwar reconstructions. The fear, the official said, was that such people would offer pessimistic scenarios, which would challenge Rumsfeld's aversion to using troops as peacekeepers; if leaked, these scenarios might dampen public enthusiasm for the war."[16]

Many symptoms of the Bush administration's understanding of power and responsibility are condensed in this episode. There is the prepon-

---

[15]Pollack, "Mourning After."

[16]George Packer, "War After the War: Letter from Baghdad," *New Yorker*, November 24, 2003.

derance of wish over fact, ideology over reality; Chalabi's claims seemed true because they matched what the administration hoped. Moreover, Bush's decisions were seldom the result of vigorous debate among his top advisers; he would not so much weigh alternatives as embrace the "strongest" option in front of him. The sort of bureaucratic gamesmanship seen in the maneuvers of the Defense Department against the State Department began immediately after September 11. The hardliners in the administration were practiced at ignoring the views of other officials, even withholding important plans from them. According to a *New York Times* report (October 24, 2004), Dick Cheney was "a driving force" behind the military tribunals created to circumvent the Geneva Conventions and give the president absolute authority over the treatment of captured terrorists. A team of White House lawyers worked on formulating the policy, and "the plan was considered so sensitive that senior White House officials kept its final details secret from the president's national security adviser, Condoleeza Rice, and the secretary of state, Colin L. Powell." Cheney himself reportedly "advocated withholding the draft from" them.[17]

As for the misjudgments that lay behind the occupation itself, they can be understood in light of Pollack's original analysis of the postwar options in *The Threatening Storm*. He identified two possible approaches to the postwar in Iraq, which he dubbed the pragmatic and the reconstruction approaches. The pragmatic approach would consider the process of "building a new, stable Iraq" far too costly and seek instead to create, in a short period of time, a "reasonably stable" political structure "regardless of whether it was equitable"; at best there would emerge "some sort of oligarchic government based on power-sharing arrangements" among Sunnis, Shia, and Kurds, and at worst "a new dictator would probably emerge . . . who would likely be someone with whom the United States could work."[18] The reconstruction approach would assume the far greater ambition of establishing the conditions of a

---

[17]For a detailed account of the role of Cheney's aide David S. Addington in circumventing potential adversaries while crafting the most far-reaching violations of international and domestic laws when it came to Guantánamo, military commissions, torture, extraordinary rendtion, and the rest, see Jane Mayer, "The Hidden Power: The Legal Mind behind the White House's War on Terror," *New Yorker,* July 3, 2006.

[18]Pollack, *The Threatening Storm,* p. 388.

democratic Iraq. Such a task was daunting and would require, Pollack estimated, 250,000 to 300,000 American troops to secure the country and a commitment of several years and massive sums of money. This assessment was based on "the belief that the current Iraqi political and social framework cannot produce a government that is stable or legitimate (and that the absence of legitimacy would inevitably contribute to its instability). This had been the pattern of Iraqi politics since independence in 1932. Only Saddam Hussein created a sort of stability, but he did so at the price of mass killing, totalitarianism, constant warfare, and genocide."[19]

The disaster of the American occupation arose, it seems to me, from a dangerous compromise between these two approaches. The minimalism of the so-called pragmatic approach fit very well with Donald Rumsfeld's outlook on military force and nation building. He crafted, in collaboration with General Tommy Franks, a quite brilliant battle plan to overthrow Saddam Hussein. It featured extensive use of special forces and a relatively small ground force that was highly mobile and interacted with devastatingly accurate air power. The plan was all the more remarkable for being created wholecloth in a very short period of time and in complete secrecy.[20] Concern for the postwar occupation was sacrificed to two of Rumsfeld's deeply held convictions. First, he was out to prove that the Powell doctrine's dependence on large numbers of troops was obsolete; he had no interest in problems that might require 300,000 soldiers or more. Second, Rumsfeld had long been a hardcore skeptic of nation building and viewed military force as something to use to crush an enemy and then go home; if the enemy eventually rises again, go back and crush them again. He turned a deaf ear to the military's own concerns that Iraq could not be secured with less than 250,000 troops, perhaps 500,000, dismissing the generals' calculation as another sign of the Powell-built military's reluctance to go to war under any circumstance.[21]

While Rumsfeld never showed the least interest in the idealistic aim of democratizing Iraq, his neoconservative brain trust at the Defense

[19]Ibid., p. 392.

[20]Cf. Woodward, *Plan of Attack*, pp. 1–179.

[21]See Michael R. Gordon and General Bernard E. Trainor, *Cobra II: The Inside Story of the Invasion and Occupation of Iraq* (New York: Pantheon, 2006), pp. 138–63 and 457–79.

Department had long made it their supreme aim. Rumsfeld's abject negligence was thus complemented by the neoconservatives' romanticism, for they truly believed that Iraqis would celebrate their invader-liberators' arrival and immediately set up a pro-Western government with democratic aspirations under a leader of America's choosing. In speeches prior to the invasion, like "Beyond Nation Building" delivered at the Intrepid Sea-Air-Space Museum on February 14, 2003, Rumsfeld even claimed that his military strategy of high-tech weapons and fewer troops would foster self-determination and proved that "America was not interested in conquest or colonization," using Afghanistan as the model: "Afghanistan belongs to the Afghans. The objective is not to engage in what some call nation building. Rather it's to try to help the Afghans so they can build their own nation." Whether miscalculation or cyncism was at the heart of Rumsfeld's battle plan for Iraq, he and the neoconservatives created an untenable synthesis of the two opposing approaches outlined by Pollack. The postwar occupation had the *aims* of the "reconstruction approach" and the *means* of the "pragmatic approach." At once grandiose and meager, revolutionary and unplanned, it was doomed from the beginning.

After the initial invasion, Iraqis needed protection from crime, looting, militias, and loyalist rebels. The Pentagon had not brought enough troops to do that, and the troops it had were untrained as police and peacekeepers. The failure to establish security had the gravest consequences and ramifications. Here was the true Hobbesian moment in Iraq. When the American occupier failed to secure a civil order, it lost all legitimacy as the temporary placeholder of Iraqi sovereignty. Where there is no secure rule of law, people will seek their own security and self-defense in any way they can. Hobbes imagined the war of all against all in a proto-bourgeois, individualistic manner; in Iraq it took the form of armed militias and fluid allegiances as people in fear for their lives and families sought security wherever they could find it. Rumsfeld and Bush loudly denounced the lawlessness of Iraqis, thereby ignoring the Hobbesian truth that in the absence of a sovereign power lawlessness is merely everyone's natural right. In less mythical, more Weberian terms, the modern state holds a monopoly on legitimate violence; without the monopoly, no legitimacy. The American occupier could not but appear illegitimate: it failed to protect civilians or property; it disbanded the

Iraqi army, leaving hundreds of thousands of men armed and unemployed; it permitted private foreign "security" personnel, otherwise known as mercenaries, to roam the country indistinguishable from the occupier (whose officials were guarded by the mercenaries rather than U.S. troops!). The occupation, returning to Hobbesian terms, simply *created* civil war.

I have heard what the talkers were talking, the talk of
                              the beginning and the end,
But I do not talk of the beginning or the end.

There was never any more inception than there is now,
Nor any more youth or age than there is now,
And will never be any more perfection than there is now,
Nor any more heaven or hell than there is now.
                    —WALT WHITMAN, *Song of Myself*

THE ORDEAL OF UNIVERSALISM

Democracy and War

Political thinkers often aspire to harmonize the principles of foreign policy with their conception of democracy. Neoconservatives, for example,
hold that a democracy's military might is a neutral instrument for keeping the nation's liberal order secure in an insecure world, and they have
advocated the invasion of foreign countries to overthrow tyranny and
create the conditions for democracy by force. The United States's own
liberal principles are thus seamlessly tied to the use of force with the
good intention of fostering liberty abroad. Robert Kagan will therefore
speak of the "uniquely American form of universalistic nationalism."[1]
For Jürgen Habermas, by contrast, universalism and nationalism are
strictly incompatible. He passionately insists that a viable vision of international order must be one that extrapolates from the workings of
democracy to global civil society. Just as human rights are as valid on the
global as on the national scene, international conflicts ought to be confronted and adjudicated in the same way that law-governed democratic
states confront and adjudicate domestic conflict and lawbreaking. This
kind of universalism Habermas calls cosmopolitanism.

[1] Robert Kagan, *Of Paradise and Power: America and Europe in the New World Order* (New
York: Alfred A. Knopf, 2003), p. 76.

These two universalisms, the neoconservative and the Habermasian, seek to transcend the traditional Realpolitik according to which every country's national interests are a kind of opaque, intrinsically justified (and morally neutral) motive for its actions in regard to every other nation. The neoconservatives draw on the deeply embedded tradition of American exceptionalism to idealize America's national-interest motives as universal in contrast to every other nation's, while Habermas evokes a cosmopolitanism rooted in Kant and newly flowered in the European Union to postulate a universal interest in rights and justice that transcends any particular nation's interests.

Although I will not advocate the Realpolitik of national interests as such, neither national interests nor Realpolitik can be a priori excluded from diplomatic decision. It is extremely rare that any country's foreign policy knowingly goes against its national interests. It indeed remains a measure of a leader's responsibility that he or she not harm the national interests. The kernel of truth in Machiavelli lies just there. The prince's ultimate responsibility is to assure the continued existence of the republic. When Max Weber ponders the spiritual consequences of the politician's commitment to the ethic of responsibility, he warns that "he who seeks the salvation of the soul, of his own and of others, should not seek it along the avenue of politics," and he is reminded of Machiavelli on this discrepancy between saving the polis and saving the soul: "Machiavelli in a beautiful passage, if I am not mistaken, of the *History of Florence,* has one of his heroes praise those citizens who deemed the greatness of their native city higher than the salvation of their souls."[2] As for Realpolitik, no one can truly rule out making compromises like allying with an unsavory regime or dispensing aid inequitably when it helps achieve a more urgent goal.

Beyond these practical limits associated with the Machiavellian moment, the very aspiration to harmonize domestic and foreign policy by deriving them from the same principles does not hold up on theoretical grounds. Political leaders obviously pursue domestic and foreign policy at the same time and sometimes provide common themes for the two. It cannot, however, be presupposed that the web of actual connections

---

[2]Max Weber, "Politics as Vocation," in *From Max Weber: Essays in Sociology,* ed. and trans. H.H. Gerth and C. Wright Mills (New York: Oxford University Press, 1946), p. 126.

between domestic politics and foreign policy is or should be the result of a unified principle. Since the internal workings of a nation are never unaffected by its external relations, it also cannot be presupposed that internal politics shape external politics more than the other way around.

There are, for example, societies whose internal development has been what the sociologist Hans Joas calls a kind of "defensive modernization," that is, their domestic modernization has been propelled by their precarious or threatening external relations to other countries. The society's "modernity" is thus a product of international power relations and possible war as much as internal forces and values. In *War and Modernity*, Joas offers a collection of penetrating studies on the tendency of modern sociology to conceive of *society* as though it were unrelated to the *state's* relations with other states. The sociological conception at once presupposes that "society" exists within the boundaries of the nation-state and then promptly ignores the state's international engagements when analyzing the society's internal workings: "All the alleged regular laws of development thus refer covertly to the reality of a state whose territories are clearly delineated, which is bounded by a body of law and administered in a modern manner, whereas the dynamics of the relationships between these states is regarded as a purely historical contingency and otherwise hardly warrants interest. Consequently, such an approach cannot adequately thematize either the particular internal characteristics of a nation-state as opposed to other historical structures, or the dependence of intra-societal processes on global, economic, political, and military processes."[3] Domestic and foreign policy shape one another, but not necessarily in the manner the neoconservatives or Habermas proclaim.

Another salient point about the effect of domestic politics and foreign policy on one another is made by Michael Walzer in a discussion of what is at issue in judging a government's actions in supposedly humanitarian interventions: "The judgments we make . . . don't hang on the fact that considerations other than humanity figured in the government's plans, or even on the fact that humanity was not the chief consideration. I don't know if it ever is, *and measurement is especially*

---

[3]Hans Joas, *War and Modernity*, trans. Rodney Livingston (Cambridge, UK: Polity, 2003), pp. 126–27.

difficult in liberal democracy where the mixed motives of the government reflect the pluralism of the society [my italics]."[4] Walzer's remark is a reminder that the very plurality of interests and vantage points that a democracy's foreign policy typically has to accommodate makes it unlikely that the theoretical search for a unified principle in foreign policy will succeed. In a sense, such a search runs contrary to the values and practices of democracy itself! Foreign policy is more likely to be forged from an ad hoc amalgam of beliefs and interests than a consistent preestablished principle.

Beyond this sort of plurality of interests, a modern democracy is pluralistic in a more fundamental sense; it inevitably draws on three contradictory frameworks: *liberal, civic,* and *social* democracy. Liberal and civic democracy differ fundamentally over what constitutes freedom. Liberal values hinge on the individual's freedom from unnecessary constraint by government, that is, what Isaiah Berlin called "negative liberty." Civic (or republican) values stress the "positive liberty" of self-rule; freedom in this sense stems from participation in self-government. Social democracy, in using government policy to direct or adjust the effects of the economic system, can clash or gel with either of these contrary understandings of freedom. Welfare policies, for example, can be thought to weaken the self-reliant individual's negative liberty or, on the contrary, to protect it by leveling the playing field on which individuals can then compete; the same policies might be criticized from the civic democratic standpoint because they transform the citizen into a mere client of the state, or conversely they might be defended for giving individuals the minimal freedom from want that is required for a citizen to make independent decisions.

Three caveats, in sum, should guide a skeptical examination of various attempts in the current global crisis to derive foreign policy from the internal principles of the democratic state. First, the modern state's inner workings are themselves often affected and shaped by the pressures of international relations. Second, the conflicts and compromises within the body politic affect foreign policy, particularly at dramatic moments of risk and decision regarding interventions and war. Third, the inner

---

[4]Michael Walzer, *Just and Unjust Wars: A Moral Argument with Historical Illustrations,* 3d ed. (New York: Basic Books, 2000 [1977]), p. 104.

workings of modern democracy are driven by the perpetual rivalry of the co-existing liberal, civic, and social frameworks, none of which is capable of giving democracy a conceptual foundation that absorbs or eliminates the other two. Since open-ended conflict over the principles of democracy is itself a principle of modern democracy, a nation never merely extrapolates its foreign policy from its democratic principles in themselves.

Postnational Cosmopolitanism versus Liberal Nationalism?

As Jürgen Habermas surveyed the dispute within the Atlantic alliance over the war in Iraq, he drew the divisions in a striking way. He saw a geopolitical divide between the "continental European powers" and the "Anglo-Saxon powers," and he saw an ideological divide between "Kantian cosmopolitanism" and "liberal nationalism." The dispute, he argued, brought out a latent antagonism that could already be felt in the Kosovo intervention: "Ever since April 1999, one saw a remarkable difference in the strategies of justification on the part of the continental European powers and the Anglo-Saxon powers. While one side was striving to draw a lesson from the tragedy of Srbrenica by shrinking the gap between effectiveness and legitimacy through armed intervention, the other saw a sufficient normative objective in the act of extending the Western liberal order, even beyond our own borders and by force if necessary."[5]

A few months after September 11, and well before the diplomatic rift over Iraq, Habermas similarly expressed his concern over the "growing dissent within the Western camp between the Anglo-Saxon and the continental countries. The former draw their inspiration from the 'realistic school' of international relations while the latter favor a normative legitimation and a gradual transformation of international law into a cosmopolitan order."[6] The idea that a cosmopolitan order ought to supplant the mere regulation of relations between states runs throughout

---

[5]Jürgen Habermas, "Le statue et les révolutionaires," trans. Christian Bouchindhomme, *Le Monde,* May 5, 2003. (My translation from the French.)

[6]"Fundamentalism and Terror—A Dialogue with Jürgen Habermas," in Giovanna Borradori, *Philosophy in a Time of Terror: Dialogues with Jürgen Habermas and Jacques Derrida* (Chicago: University of Chicago Press, 2003), p. 40.

Habermas's post–Cold War writings. Even when he acknowledges how limited the signs of a cosmopolitan order are, he speaks of transition and transformation: "The contemporary world situation can be understood at best as a transitional stage between international and cosmopolitan law. But many indications seem to point instead to a regression to nationalism."[7] Habermas's historical view is premised on the idea of a *developmental sequence:* nationalism / international law / cosmopolitanism. International law arose to regulate the relations among competitive, often belligerent nations, but it has only partially tamed nationalism and limited wars. Beyond it lies the cosmopolitan order in which human rights and international decision making within a global legal framework would guarantee human rights and establish peaceful relations among states.

Cosmopolitanism, in Habermas's revision of Kant, hinges on the idea that the rule of law as it is achieved within the constitutional state should be the model of a cosmopolitan order among states: "Cosmopolitan law is a logical consequence of the idea of the constitutive rule of law. It establishes for the first time a symmetry between the juridification of social and political relations both within and beyond the state's borders."[8] Such a vision represents a strong version of the aspiration to harmonize democracy and diplomacy within a single theoretical construct: it calls for a *symmetry* between domestic and global law. The vision also exhibits Habermas's habit of organizing normative frameworks into developmental sequences: cosmopolitan law becomes the *logical consequence* and *progressive development* of the constitutional rule of law within democratic states.

The cosmopolitan perspective draws on the experience of European integration. The nation-states of Europe have overcome their history of armed conflict by creating a postnational politics, first on the economic and administrative planes, increasingly on the level of elections and gov-

<hr />

[7] Jürgen Habermas, "Kant's Idea of Perpetual Peace: At Two Hundred Years' Historical Remove," in *The Inclusion of the Other: Studies in Political Theory,* ed. Ciaran Cronin and Pablo De Greiff (Cambridge: MIT Press, 1998), p. 183. When Habermas wrote these lines in 1995 he had in mind the conflicts in the Balkans and in the former Soviet Union; there is every indication that he would today consider American unilateralism in the war in Iraq to exemplify the regression to nationalism he fears as a setback to cosmopolitanism, even to international law.

[8] Ibid., p. 199.

ernment, and ultimately, it is hoped, in diplomacy and international re-lations. The very existence of the European Union itself shows that na-tion-states with a catastrophic history of imperialism and war, revolu-tion and genocide, can attain a new order among themselves that more and more resembles the *domestic* security of a single nation. It has not, however, been altogether clear, certainly not so clear as Habermas sug-gests with his distinction between "continental" and "Anglo-Saxon" worldviews, whether Europeans themselves are really committed to a unified European foreign policy or to a cosmopolitan model for con-ducting such a policy. When the United States and Britain decided to invade Iraq without UN approval in 2003, they garnered support from several continental governments, including Spain, Italy, Poland, and Hungary. The crisis provoked in May 2005, when the French, whose leaders have long been the most vocal advocates of a European foreign policy, voted against the proposed European Constitution in a referen-dum, further postponed any semblance of a unified European stance. That said, Habermas is not wrong when he characterizes the aspiration raised by European Union's historic achievement in the following terms: "Cosmopolitans see a federal European state as a point of departure for the development of a transnational network of regimes that together could pursue a world domestic policy, even in the absence of a world government."[9]

A *world domestic policy* refers to the idea that the lawful regulation—and policing—of international relations could attain the essential fea-tures of domestic security. In place of the ever present possibility of war between nations there could arise a *global* rule of law upheld by the "transnational network of regimes." Europe in the cosmopolitan's eyes aspires to be the model, the architect, and a prime enforcer of this new world order. With Kant's warning about the potential despotism of world government in mind, Habermas does not advocate a global su-perstate but imagines instead that Europe and other Europe-inspired federations would expand far enough that, in conjunction with a revi-talized United Nations entrusted with codifying the laws and legitimat-ing their enforcement via sanctions, embargoes, or interventions when required, the world could so to speak police itself.

[9]Jürgen Habermas, "Toward a Cosmopolitan Europe," trans. Max Pensky, *Journal of De-mocracy* 14, no. 4 (October 2003), p. 96.

Kant with Hobbes

The search for a global domestic policy is fraught with pathos in the drama of Habermas's thought. The internal intellectual struggle is palpable. Empirical lucidity grinds against normative hope, sounding a note of despair at the very moments that cosmopolitanism is affirmed: "Surely, everyone today is in agreement that the idea of a just and peaceful cosmopolitan order lacks any historical and philosophical support. But what other choice do we have, besides at least striving for its realization?"[10]

Compare this remark to the passage it echoes in Kant's 1795 essay "Perpetual Peace": "And while the likelihood of [perpetual peace] being attained is not sufficient to enable us to *prophesy* the future theoretically, it is enough for practical purposes. It makes it our duty to work our way toward this goal, which is more than an empty chimera."[11]

Why does Kant seem less despairing than Habermas? Why does Kant see an arduous duty where Habermas sees a last resort?

There are of course the historical differences. The devastation of modern war, especially the targeting of civilians and the development of nuclear weapons, magnifies every shortcoming in the contemporary world order. There is, however, another difference that is crucial to understanding Kant. His own cautious pessimism comes from his sense of the ineradicable evil and violence at the heart of human nature. Exactly contrary to what today's purveyors of the Kant-versus-Hobbes theme suggest, Kant holds a darker view than Hobbes. Hobbes starts from the idea that every individual has the natural right to protect his own existence and from there postulates the hypothetical state of nature, where in the absence of a civil order there would simply be "the Warre of every one against every one." Conversely, civil order originates in the covenant through which the multitude forgoes violence one against another by instituting a sovereignty that reigns over all. Hobbes thought of the re-

[10]Jürgen Habermas, "The Gulf War: Catalyst for a New German Normalcy?" in *The Past as Future* (Interviews with Michael Haller), ed. and trans. Max Pensky (Lincoln: University of Nebraska Press, 1994), p. 22.

[11]Immanuel Kant, "Perpetual Peace: A Philosophical Sketch," in *Kant's Political Writings*, ed. Hans Reiss and trans. H. B. Nesbit (Cambridge: Cambridge University Press, 1970), p. 114.

lation between nations as *less* insecure than the state of nature conjured up by civil strife, since the sovereign's very existence is the product of the covenant made *by* the multitude to establish *their own* peace and security. The sovereign makes the decision for or against war for the sake of the multitude's peace and security: "And because the End of this Institution, is the Peace and Defence of them all; and whosoever has right to the End, has right to the Means; it belongeth of Right, to whatsoever Man, or Assembly that hath the Soveraignty, to be Judge both of the meanes of Peace and Defence; and also of the hindrances, and disturbances of the same; and to do whatsoever he shall think necessary to be done, both before hand, for the preserving of Peace and Security, by prevention of Discord at home and Hostility from abroad; and, when Peace and Security are lost, for the recovery of the same."[12] For Hobbes, the relative prudence of states in their relations with one another followed from the sovereign's inherent concern for the commonweal. The sovereign's absolute power over each and every individual in the body politic is at the same time his own empowerment *in* and *as* the commonwealth of all.

While Hobbes forged his views horrified by civil war, Kant forged his in response to the horrors of wars between states: "Peoples who have grouped themselves into nation states may be judged in the same way as individual men living in a state of nature, independent of external laws; for they are a standing offence to one another by the very fact that they are neighbors."[13] The state of nature between individuals was for Kant hypothetical, but the state of nature between nations was frighteningly real. The Kantian heart of darkness lies in how war unleashes the worst in human nature: "Although it is largely concealed by governmental constraints in law-governed civil society, the depravity of human nature is displayed without disguise in the unrestricted relations which obtain between the various nations."[14] To protect against this naked depravity, nations ought to undertake a pacification of their relations with one another comparable to the Hobbesian transition from the state of nature to a commonwealth. Kant's appeal for a "federation of free states" is

[12]Thomas Hobbes, *Leviathan,* ed. C. B. Macpherson (New York: Penguin, 1968), pp. 232–33 [ch. 18].
[13]Kant, "Perpetual Peace," p. 102.
[14]Ibid., p. 103.

grounded just there: "Each nation, for the sake of its own security, can and ought to demand of the others that they should enter along with it into a constitution, similar to the civil one, within which the rights of each could be secured."[15]

Habermas's Agon *with Schmitt*

When Habermas, honoring the bicentennial of "Perpetual Peace," undertakes to revaluate and revise Kant in light of contemporary international conditions, he chooses as the polemical *agon* of his argument Carl Schmitt. He apparently considers Schmitt's challenge to cosmopolitanism to be so fundamental that he links the reformulation of Kant directly to the refutation of Schmitt. Defeating Schmitt becomes central to salvaging Kant. (The burden of this double argument is, I will suggest, too heavy to bear—and, besides, it is unnecessary.) Schmitt denounced any politics that acted on behalf of humanity in international affairs, and he specifically disputed the legitimacy of human rights. The purposes and motives of his arguments were thoroughly disreputable: early in his career he vehemently denounced the American aim of making the world safe for democracy in World War I, the Versailles Treaty, and the Wilsonian League of Nations; as an intellectual star during the Third Reich, he defended the legality of Hitler's dictatorship; and, finally, he objected in the postwar to the prosecution of Nazis on the grounds that the very concept of war crimes is illegitimate. And yet the purport of Schmitt's arguments is often forceful and continues to find adherents on the left as well as the right. He warned that when states go to war claiming that their enemy is a threat to humanity, they in effect destroy the rule-governed behavior between states at war, transform the political coherence of armed conflict into an inchoate moral crusade, and end up licensing unrestrained violence against the dehumanized enemy of humanity and thereby turn themselves into inhumane defenders of humanity. The red thread of these arguments is that the introduction of the morality of good and evil into the politics of friend and foe destroys politics by moralizing it, and that when politics is thus transformed into morality there is no limit to the hatreds and injustices that follow.

[15]Ibid., p. 102.

Habermas picks up Schmitt's challenge and rebuts the idea that human rights represent a moralization of politics. Human rights, he argues, are fundamentally *political* rights, even though they have a *moral* aspect: "The concept of human rights does not have its origins in morality, but rather bears the imprint of the modern concept of individual liberties, hence of a specifically juridical concept. Human rights are juridical *by their very nature.* What lends them the appearance of moral rights is not their content . . . , but rather their mode of validity, which points beyond the legal orders of nation-states."[16] I understand Habermas's claim as follows: On the one hand, human rights are political rights since they originate in liberal thought and modern constitutions as individual rights vis-à-vis the state; and, on the other hand, they have the feel of moral rights since they are valid even for individuals who do not live under a rights-granting constitution and rights-protecting state; *therefore,* this universal validity of politico-legal rights puts on the historical horizon a legal and political order beyond nation-states.

The conclusion—the *therefore*—is something of a leap. Human rights can certainly be said to originate in the political realm. When those who create a shared polity as citizens endow themselves with rights, there are some rights that they extend to noncitizens. These are the core of what is understood, or comes to be understood, as *human* rights, since they are individual rights tied to being a person, not a citizen. There is thus an elegant paradox at the origin of human rights: although human rights derive from civic rights, the citizens who invent civic and human rights consider *human* rights to be, precisely, more basic and fundamental than civic rights. The scope of human rights— that is, how widely do they apply?—adds another complexity. The universalism of human rights has two thresholds. The first threshold, *these are the rights due to the noncitizens of this polity,* leads to a second: *these rights are due to the citizens or noncitizens of any polity.* Habermas is too quick to dispel the complexity and paradox according to which citizens must create civic rights *before* creating the *more fundamental* human rights. In his view, since human rights *originate* in the political realm and apply to individuals *universally,* they *ultimately* call for a *global* political order. The *arche:* universal rights. The *telos:* a global political order.

[16]Habermas, "Kant's Idea of Perpetual Peace," p. 190.

There is also in Habermas's argument an obvious overstatement in eliminating all moral *content* from human rights. Such a view cannot account for either the origins of human rights or the reasons why the violation of human rights beyond our own borders and culture can so disturb us. It is arbitrary to bracket the ethical and religious traditions that feed into the modern political genealogy of rights. For example, the fundamental human rights founded on the integrity of the person—from the right to "life and liberty" to the prohibition against torture—contain, as part of their *content*, values and symbols from ethical and religious traditions. *Your body is a temple* is a source, filtered through Westen culture, of our understanding of, and our feeling for, the rightness of rights. The rights are indeed political, but their content and origin are not purely political.

So, too, the agonized response that can well up within public opinion in response to reports and images of human rights violations is moral in a more basic way than Habermas admits. For him, the moral dimension is simply by analogy: "human rights . . . ground rights for all persons and not merely for citizens. . . . It is this universal range of application, which refers to human beings as such, that basic rights share with moral norms . . . [and] lends [human rights] the appearance of moral rights."[17] It seems to me the situation is quite different. When we witness human rights abuses, there are strong moral feelings, including feelings of guilt that intensify the more impossible an effective intervention seems. The gap between the rights we cherish from within our own polity and the outright violation of those rights witnessed elsewhere is an open wound in the symbolic flesh of humanity, and the sight of this wound stirs the moral response to seek some political remedy. However, the nature of the political remedy is intrinsically uncertain and ungeneralizable, since not every regime abusing human rights can be dealt with effectively or in the same way. Such inequities exacerbate the guilt and the wound.

Habermas looks to jump over the conundrum of moral outrage and effective action, at least theoretically, by concluding that the very existence of a gap between rights-respecting and rights-violating states logically points to the necessity of a global (cosmopolitan) rule of law. To

[17]Ibid., pp. 190–91.

get to that conclusion, however, he has had to deny the paradox and complexity of human rights and the moral nature of the torment that awakens the will to intervene in the first place. He arrives at this unfortunate juncture because he is convinced that the only way to refute Schmitt's rejection of human rights and cosmopolitanism is to subordinate the moral dimension of human rights completely within the political frame.

There is a different argument to made against Schmitt that is just as effective and far less drastic. First, let us straightaway accept that Schmitt has accurately identified the *dangers* inherent in the prosecution of war crimes and in humanitarian interventions through the use of force. Such prosecutions and interventions run all the risks of vengeful excess inherent in righteousness. However, they are not therefore, as Schmitt claims, invalid in principle. At issue in international relations, rather, is how to establish the procedures and criteria that put judicious limits on the prosecutions and interventions which override state sovereignty in order to prevent, stop, or punish those acts of a regime that "shock the moral conscience of mankind."

I borrow this last phrase from Michael Walzer, who uses it to give a criterion for armed intervention on the part of a nation or coalition to stop massacres, as in his example of India's intervention against Pakistan on behalf of Bengalis in 1971. His remarks bear directly on two of the problems I see in Habermas's understanding of humanitarian intervention, namely, where to place the moral dimension and how to measure international legitimacy:

> No doubt, the massacres were a matter of universal interest, but only India interested itself in them. The case was formally carried to the United Nations, but no action followed. Nor is it clear to me that action undertaken by the UN, or by a coalition of powers, would necessarily have had a moral quality superior to that of the Indian attack. . . .
>
> Humanitarian intervention is justified when it is a response (with reasonable expectation of success) to acts "that shock the moral conscience of mankind." The old-fashioned language seems to me just right. It is not the conscience of political leaders that one refers to in such cases. . . . The reference is to the moral convictions of ordinary

men and women, acquired in the course of their everyday activities. And given that one can make a persuasive argument in terms of those convictions, I don't think that there is any moral reason to adopt the posture of passivity that might be called waiting for the UN (waiting for the universal state, waiting for the messiah . . .).[18]

Walzer sets the standard for armed intervention quite high, since a shock to the moral conscience of mankind is sharply distinct from the intervening country's own national interests and designs. At the same time, those interests and designs do not have to be absent in practice or neutralized in theory for the intervention to be justified; what counts is that what prompts the taking of action be the universalizing moral response of the public.

Is such a response susceptible to error and manipulation? Of course it is, as moral responses and political judgments always are. Walzer's position has the strength of admitting the fallibility of moral and political judgment, as well as the presence of national interests and designs, *without* abandoning the task of establishing principles for humanitarian interventions. This reveals the decisive weaknesses in Habermas's approach. In his *agon* with the American neoconservatives, he seeks a juridical-political framework in international relations that would exclude or altogether transcend national interests and Realpolitik. In his *agon* with Carl Schmitt, he seeks a framework that would preclude righteousness and vengeance in the arrest and prosecution of war criminals. Can any policy or principle truly preclude the irrational elements in the motives of actors on the international scene? Is that even desirable? Political and juridical frameworks must be capacious not confining when it comes to human motive. They must steer the irrational into an arena of reasoned discussion, deliberation, and decision, not attempt to expunge it from the processes of judgment.

The contrast of Walzer's position and Habermas's also helps clarify the question I have posed regarding the supposed transition from international law to cosmopolitan order. Walzer argues that an intervention's legitimacy—whether moral, political, or legal—does not intrinsically require that the decision to intervene be made under UN or even trans-

[18]Walzer, *Just and Unjust Wars*, p. 107.

national auspices. Habermas lamented that the situation in the Security Council during the crisis in Yugoslavia left the task of intervening to European countries, the United States, and NATO. For him, the absence of UN authorization lessened the legitimacy of the intervention and enhanced the tension between continental European and Anglo-American approaches. Walzer makes the point in his preface to the third edition of *Just and Unjust Wars* in 2000 that while the "wider consensus" of a UN intervention might have several advantages, the United Nations is no more immune from bad judgments than individual nations, and, second, there are occasions when a neighboring (and therefore *interested*) country is better situated to assess and act: "Even a global regime with a global army, however, would sometimes fail to act forcefully in the right place at the right time. And then the question would arise whether anyone else, in practice any state or alliance of states, could act legitimately in its place. Humanitarian interventions like those in Cambodia [by Vietnam to close down the 'killing fields'] or Uganda [by Tanzania to overthrow Idi Amin], which never would have been approved by the UN, might have been impossible had the UN actually disapproved, that is, voted against them. There are obvious disadvantages to relying on a single global agent."[19]

In the international arena today, the elements of a cosmopolitan order that are associated with UN action exist side by side with the unilateral and multilateral actions of interested, self-interested, complexly motivated nation-states. This situation is not a *transition* from international law to cosmopolitan order but an *amalgam* of national, international, and cosmopolitan elements. It is better to embrace the mixed international-cosmopolitan situation and proceed to clarify the guiding principles of humanitarian intervention rather than paint oneself into a corner where intervention cannot be justified (legitimated) except by an as yet unattained, and perhaps unattainable, cosmopolitan order. So, too, it is better to embrace, unlike either Schmitt or Habermas, the mix of morality and politics in the concept of human rights and proceed to detail the inherent dangers of moralizing politics rather than deny that the feelings and motives behind humanitarian intervention are moral.

[19]Ibid., p. xiv.

Hobbes with Kant

In making his argument against Schmitt, Habermas intends simultaneously to take Kantian cosmopolitanism beyond its eighteenth-century horizon: "Because Kant regarded the bounds of national sovereignty as inviolable, he conceived of the cosmopolitan community as a federation of states, not of world citizens. This was inconsistent in that Kant derived every legal order, and not just that within the state, from the original right that attaches to every person 'qua human being.' Every individual has a right to equal liberties under universal laws ('since everyone decides for everyone and each decides for himself')."[20] Habermas claims to eliminate inconsistency from Kant's conception of cosmopolitanism. The argument parallels his refutation of Schmitt and might be paraphrased as follows: Since human rights derive from a postulated legal order applied to individuals universally, and since the existing international legal order, premised as it is on the relations among sovereign states, does not provide such a legal order, the universal human rights of individuals *therefore* require a post–nation-state, postinternational legal order.

The *therefore* is as big a leap in the beyond-Kant argument as in the anti-Schmitt argument. It can seem logical enough: cosmopolitan rights require a cosmopolitan order. In the anti-Schmitt argument, Habermas's search for consistency leads him to ignore two essential features of human rights: first, the elegant paradox that (prepolitical) human rights are created by a political act, and, second, the ineradicable tension between the moral and political elements of human rights. As regards Kant, Habermas overstates the supposed inconsistency in Kant's failure to point beyond the legal and political order of actual states. Kant had very good reasons for not looking beyond national sovereignty, since his era was just undertaking the experiment with popular sovereignty and witnessing the emergence of democratic states that needed protection from foreign, counterrevolutionary powers. Kant is perhaps not inconsistent at all. Rather, what can be seen here is that for all the rigor and abstraction of the universal maxims he puts

[20]Habermas, "Kant's Idea of Perpetual Peace," p. 180.

forth to define right, freedom, duty, and equality, he nevertheless keeps his normative claims within the orbit of the art of the possible. Therein lies a further reason why he insists that philosophy cannot prophesy perpetual peace.

For Kant, world government was the only conceivable step beyond nation-states, international law, and a "federation of free states," and he believed world government would inevitably be prone to despotism. Habermas acknowledges Kant's wariness and disclaims that the enforcement of cosmopolitan right would require a global superstate. The disclaimer strikes me as an equivocation. If the cosmopolitan order is to be a legal order enforced by a "transnational network of regimes" under UN auspices, it surely resembles world government; if it does not resemble world government, then it falls far short of a "legal order" in the sense Habermas means. The equivocation scarcely veils the fact that Habermas, unlike Kant, has not measured his normative claims against the political art of the possible.

When political principle confronts politics, what is required is not consistency so much as a tolerance for paradox. Negative Capability—that is, being "capable of being in uncertainties, Mysteries, doubts, without any irritable reaching after fact & reason," as Keats defined it—is a virtue for the political theorist just as much as the poet. The paradox at issue lies in the task of *relativizing sovereignty*. For that is the question posed by the very idea of humanitarian intervention. Under what conditions can, and should, a state's sovereignty be violated, suspended, or overridden? In recent years Gareth Evans, former Australian foreign minister and now head of the NGO International Crisis Group, has spearheaded efforts to redefine sovereignty so as to give humanitarian interventions a more cogent foundation. Sovereignty is traditionally defined primarily or solely as a nation's right not to have other nations interfere in its internal affairs. The revisionists argue that the definition of sovereignty should include a government's *responsibility to protect* the people over whom it is sovereign. A murderous despotic regime or a regime engaged in ethnic cleansing or genocide has, according to this definition, violated or defaulted on its own sovereignty. Other nations are justified, in some sense even obligated, to take upon themselves this responsibility to protect until it can be restored within the country's own political system. The horizon is to *restore* sovereignty.[21]

No one seems to have noticed that this admirable innovation in cosmopolitan thinking resurrects the essence of sovereignty presupposed by Hobbes! Hobbes's axiom is not Schmitt's *Sovereign is he who declares the state of exception.* Rather, Hobbes's axiom implicitly is *Sovereign is he who protects the multitude.* Could it be that Hobbes is the unacknowledged prophet of humanitarian intervention? Not directly, obviously. There is, however, a crack in the absoluteness of Hobbesian sovereignty, and this crack is revealed by Kant in the very discussion Habermas is alluding to, namely, the middle section of the great essay, "On the Common Saying: 'This May be True in Theory, but it does not Apply in Practice.'" The essay is on the face of it an inauspicious starting point for relativizing sovereignty, since Kant is arguing for the absolute prohibition on rebellion, even against tyrannical authority:

> all resistance against the supreme legislative power, all incitement of the subjects to violent expressions of discontent, all defiance which breaks out into rebellion, is the greatest and most punishable crime in a commonwealth, for it destroys its very foundations. This prohibition is *absolute*. And even if the power of the state or its agent, the head of state, has violated the original contract by authorising the government to act tyrannically, and has thereby, in the eyes of the subject, forfeited the right to legislate, the subject is still not entitled to offer counter-resistance.[22]

Having thus defended the absolute prohibition on rebellion, Kant turns around to "maintain that the people too have inalienable rights against the head of state, even if these cannot be rights of coercion." With Hobbes, Kant holds that the people have no "rights of coercion" against the sovereign, but he criticizes Hobbes's "quite terrifying" notion that "the head of state . . . can do no injustice to a citizen, but may act toward him as he please." In the eyes of any Anglo-Saxon or continental

---

[21]See Gareth Evans and Mohamed Sahnoun, "The Responsibility to Protect," *Foreign Affairs* (November/December 2002), and numerous related speeches and reports on the website of the International Crisis Group (www.crisisgroup.org).

[22]Immanuel Kant, "On the Common Saying: 'This May be True in Theory, but it does not Apply in Practice,'" in *Kant's Political Writings*, p. 81.

European political thinker for whom the American and French revolutions are watersheds in democratic inauguration and the struggle against tyranny, Kant's defense of inalienable rights seems utterly toothless, if not hypocritical, when combined with an absolute prohibition on rebellion. It is a far cry from the claim of the signers of the Declaration of Independence "that whenever any form of government becomes destructive of these ends, it is the right of the people to alter or to abolish it, and to institute new government." Habermas might better have attacked Kant's notion of sovereignty per se, since if Kant cannot countenance the people overthrowing a tyrant that oppresses them, it would certainly be inconceivable for him to grant any combination of foreign powers more legitimacy than the oppressed people themselves to intervene against tyranny.

Let us not miss, though, the crack that Kant nevertheless discloses in Hobbesian sovereignty. Kant first affirms that among human beings' inalienable rights is the right to make one's own judgments, and he then claims that a subject is therefore "entitled to make public his opinion on whatever of the ruler's measures seem to him to constitute an injustice against the commonwealth." This assertion is in keeping with the original premise that no subject may "offer counter-resistance" to the sovereign: "The non-resisting subject must be able," Kant reasons, "to assume that his ruler has no *wish* to do him injustice"; if he believes he has been done an injustice, he will therefore consider it the result of an "error or . . . ignorance of certain possible consequences of the laws which the supreme authority has made." The inalienable right to one's own judgment thus gives rise to an entitlement (Kant does not say a right) to make the opinion known. Otherwise, says Kant in his ultimate rejoinder to Hobbes, one would be granting that the sovereign "receives divine inspiration and is more than a human being."[23] Kant thus effectively counters Hobbes's refusal to grant the subject any "inalienable rights against the head of state," and he does so not simply with the reminder that the mortal God is human, not divine, but more importantly by embracing Hobbes's own implicit axiom that the sovereign is he who protects. For it is only on the basis of that axiom that the subject's inaugural assumption that "his ruler has no *wish* to do him

[23]Ibid., p. 84.

injustice" is reasonable (as the so-called original contract must be) and not absurd.

Some 350 years after Hobbes and over 200 years since Kant, the axiom that sovereignty includes the responsibility to protect is finally, tentatively, coming into its own in political discourse. Hobbes *is* the unwitting prophet of humanitarian intervention, and Kant is the Hobbesian critic of Hobbes who brings out that sovereign power is inseparable from sovereign responsibility. Neither Hobbes nor Kant could have fathomed justifying the violation of sovereignty in the name of humanitarian intervention. But it is not because either thinker is inconsistent. On the contrary, their principle, *Sovereign is he who protects the multitude,* has become belatedly relevant in our time because it has drifted into the orbit of the art of the possible. The principle demands a tolerance for the new paradox, which would simply have been nonsensical to either Hobbes or Kant, of the violation of sovereignty as a means of restoring sovereignty. The principle is now plausible rather than merely illogical for two reasons: first, the twentieth-century's experiences of genocides and tyrannies that might have been stopped have already relativized sovereignty in the "moral conscience of mankind"; and, second, post–Cold War international conditions make humanitarian interventions more feasible.

The most frequent problem today is not a lack of legal structures and international institutions but the will of nations to act on behalf even of peoples threatened with extinction, as Western and international inaction in Rwanda and later in Sudan amply proves. A population in distress from tyranny or genocide most immediately needs foreign help. But this is merely their most immediate need. What they ultimately need is a reconstructed state whose constitution and institutions protect a sufficient array of civic and human rights. Habermas's formula that cosmopolitanism lies beyond the nation-state should be reversed. It is, rather, and paradoxically, "the legal orders of the nation-state" that lie beyond cosmopolitan order, since the cosmopolitan order will always be partial, approximate, cobbled together.

The current assemblage of nation-states and cosmopolitan rights, international law and transnational alliances, an imperfect UN and self-interested Western powers, is finally a better image of the desired outcome than is Habermas's developmental model. The dream of a world domestic policy is a dubious ideal when it comes to the guarantee of basic human rights. The globe is not, and never will be, a pacified territory "policed" by a power having an *effective* monopoly on legitimate violence. Human rights can only be guaranteed by individual polities committed to sustaining and protecting rights; outside intervention can stop the violation of human rights only where intervention is economically and militarily feasible, and even then the uncertainties of military force, civil strife, and nation building are fraught with risks quite unlike domestic police actions. The only real hope is that more and more nations become constitutional, rights-protecting commonwealths with expanding democratic institutions. And the only real measure of progress in securing human and civic rights is state by state. The stricter Habermasian sense of cosmopolitan order cannot orient action because it treats the current amalgam in international affairs merely as a defect retarding the advent of the ideal. That is why Habermas, unlike Kant, is ultimately constrained to say that the cosmopolitan ideal today "lacks any historical and philosophical support."

Europe, or, the Empire of Rights

Hobbes-versus-Kant, whether in Robert Kagan's Mars-Venus theme or Habermas's opposition of Anglo-Saxon liberal nationalism and European postnational cosmopolitanism, has proved inadequate for capturing the U.S.-European difference. Yet the difference does exist, and it clearly hinges on nationhood. Americans see the nation as the center and circumference of their identity as citizens, while Europeans since World War II have considered nations as such the source of danger and insecurity and have therefore experimentally looked beyond the nation-state for new possibilities of civic identity. By the same token, however, divergences within Europe between "sovereignists" and "Europeanists" inflect all debates over a European constitution and European foreign policy. In matters of foreign policy, Americans take up the attitude of nation, while European countries see that their influence on international affairs can only be achieved through transnational endeavors. History and geography will leave this difference intact for some time. Nevertheless, many observers err in casting the U.S.-European difference as a simple polarity.

Liberal nationalism and transnational cosmopolitanism are malleable orientations, not fixed stances. They are two ongoing experiments in universalism, each of which has been born from distinct experiences of democracy and war. The American and European experiments are not polar opposites because they are not the globe's only alternatives. Francis Fukuyama's end-of-history thesis lulled many into the belief after 1989 that no political ideology would ever again challenge Western liberalism. September 11 discomforted this belief. Radical Islam is competing for the hearts and minds of Muslims, no less than a fifth of the world's population; since liberalism and democracy have so little a foothold in the Muslim Middle East, so-called moderate Islam is in a precarious position between the authoritarian regimes on the one hand and the radicals capable of stirring populist loyalty and violent commitment on the other. Consider, further, that China is expected to emerge as the next economic-military-diplomatic giant on a par with the United States and Europe. China is more thoroughly nationalistic than the United States, and it has its own blend of imperial and isolationist instincts. If Russia recovers its strength as a colossus on the world stage, most likely wielding its power through its control of huge energy resources, it too will be animated by nationalism. Neither Chinese nor Russian nationalism is liberal or universalistic. China is in the midst of a grand social and political experiment as unprecedented as Europe's. Unique among nations since 1989, China has set out to prove that a populous modern society can be capitalist and socialist at the same time and still not become democratic. What influence this model will have on other countries and what sorts of geopolitical strategies a nationalistic one-party Chinese superpower might devise to compete internationally cannot at this moment be predicted.

What the United States and Europe have in common sharply distinguishes them from Russia, China, and the Muslim world today: the yoking of democracy and capitalism, political liberalism and market economy. There are to be sure major differences in the ways the United States and Europe treat the role of the state in regulating the market and blunting the ill effects of capitalism, but while those differences are magnified in domestic debates between neoliberals and social democrats, they look much smaller when compared to nondemocratic countries, from Saudia Arabia and China to Pakistan and Russia.

The United States and Europe are not merely committed to democracy and capitalism, they are also devoted to *spreading* democracy and capitalism. The American approach to this task, with military strength playing a decisive role, has often been called "empire," from Raymond Aron's astute analysis of the *Imperial Republic* in the 1970s to Hardt and Negri's dubious visionary exposé of postmodern *Empire.* But since the fall of the Soviet empire, the power that seems to have spread democracy and capitalism most effectively is the European Union. It has expanded to include no fewer than twenty-five nations, including many Central and Eastern European countries that had been under Soviet domination. EU membership requires that a country meet stringent political and economic standards, including an array of social as well as political rights. Formally, these demands are simply rules and regulations that any country *wanting* membership must meet, but there is a coercive aspect to the negotiations. While application for membership is voluntary, exclusion from the European Union for any country that lies on its frontier would ultimately result in economic disadvantage and decline. The very existence of the European Union makes the stakes of failing to conform to its demands very high. Usually the petitioning country's more dynamic political and economic elites are especially aware of the advantages of membership for their nation as well as themselves. The EU's soft coercion imposes rules and regulations that would often be politically unattainable otherwise by the prospective member's own leaders.

It is possible, then, that this is the emergence of a new kind of empire: An empire that conquers by incentives, negotiations, and requirements. An empire that radically reforms every candidate for membership before absorbing it as a partner on equal terms. An empire that relentlessly widens its borders and extends its territory in order to spread democracy and capitalism. An empire that creates a free market among its members, a market that is at the same time carefully regulated within the terms of the Union's continual compromises between neoliberal and social-democratic policies. An empire that, most essentially, requires each newly absorbed country to grant its own citizens and resident aliens the world's most extensive set of political and social rights.

Europe is an *empire of rights.* Its experiment is so bold that many of its oldest members have balked in recent years, especially as expansion brought countries into the Union that were once considered lost to

Asiatic despotism and are now more neoliberal and pro-American than the peoples of Western Europe. The boldest overture of the European Union thus far has been to accept putting Turkey on the path to membership, setting off considerable anxiety and backlash. European expansionists, like Timothy Garton Ash, urge Europe to more energetically pursue countries that were once part of the Soviet empire, including in Central Asia, in order to counteract the autocracies and religious sectarianism that threaten to sweep the region.[24]

Why not just follow Habermas and call the European experiment *transnational cosmopolitanism?* I have no objection, so long as the term is wrenched from the meaning he attaches to it. For Habermas does not really describe the European Union. Moreover, he depletes the language of political theory by striving to liquidate or repudiate a series of notions that are deeply ingrained in Western political thought, notions that retain a relevance for politics and statecraft today, even if their interpretation has to be "twisted" with enough torque to loose them from their original contexts: namely, the notion that the human inclination to depravity is ineradicable (Kant); that the fear of death and the right to self-preservative violence are integral to political reason itself (Hobbes); that the prince's responsibility to assure the continued existence of the republic transcends, for the prince, all other moral considerations (Machiavelli); that the individual's desires and will are ultimately refractory to the ethical demands of collective life (Nietzsche, Freud, Weber); that a state's integrity, independence, legitimacy, and rule of law—that is, its sovereignty—depend at the limit on its ability to suspend the rule of law in an emergency (Schmitt).

I do not see the wisdom in seeking a political theory that would exlude all these notions. Habermas is a passionate spokesman for the value of committing oneself, above all, to respect for the rules of reasoned discussion, delight in debate, acceptance of majority decision. It is a compelling, honorable commitment. I share it. But as Max Weber understood, when one decides to make such a commitment, the decision is in itself *irrational.* The commitment to reasoned debate as supreme value is irrational. In disavowing the irrational element in his own choice of values, Habermas falls back on the idea that these partic-

---

[24]Timothy Garton Ash, "Exchange of Empires," *Guardian,* May 20, 2005.

ular values are intrinsic to the very use of language and discourse. The particular is made universal. In turn, his forceful and passionate argument on behalf of constitutional patriotism over national identity and nationalism often tilts into a quixotic denunciation of all the sorts of passions and interests that motivate political participation in the first place, as though the citizen should be motivated by the proceduralist belief in reasoned argument alone.

The limitations of a merely proceduralist Europe were abruptly exposed in 2005 when the voters of France and the Netherlands, two of the EU's original six founders, rejected the European constitution by decisive margins. France had been the driving force behind the European Union for decades. Suddenly the French balked at embracing their own creation. They did not identify with it, they did not imagine themselves acting through Europe or as Europe. The solid rejection pointed up the drawbacks of a "constitution" that is an administrative tome of several hundred pages rather than a literary-political document addressed directly to every citizen to affirm as his or her own expression and pledge. The vote also underscored problems in the French political system. President Jacques Chirac gained a false majority in 2002 when his opponent was the right-wing extremist Jean-Marie Le Pen; he then enjoyed ephemeral mass popular support for his opposition to the war in Iraq, but all the while he governed poorly; the referendum, which he himself had chosen to call for political calculations all of which turned out to be wrong, became in part a protest against his and his prime minister's leadership.

The deeper problem, though, is that Europe does not inspire even its own creators. Three decisive elements in the No vote reveal the lineaments of missing inspiration. The extreme Right and the sovereignists could not imagine how France might advance its values and glory via Europe. The segments of the Left that remain anticapitalist—whether by conviction, confusion, or opportunism—refused the constitution's double affirmation of democracy and capitalism; either they do not at bottom accept that social-democratic ideas must contend with neoliberal ideas in debates of uncertain outcome (the dogmatic option), or else they believe that such debates will be harder to win on a European than a national scale (the ostrich option). A third element, dominated by the Catholic Right and xenophobes, rejected the constitution in an anticipatory protest against Turkey's eventual membership in the European Union, because they cannot envision Europe beyond a geo-imaginary Christendom. In sum: extreme right-wing nationalists, nostalgic sovereignists, doctrinal Trotskyists and communists, opportunistic Socialists, Islamophobic Catholics, and the inevitable smattering of just plain racists formed a coalition to defeat the constitution.

This coalition of the unwilling was nothing but splinters, so ad hoc that it disappeared the day after the referendum. Not only did it lack any plausible alternative to the proposed constitution, but the various factions did not advance any of their own causes one iota by defeating the referendum: France did not become stronger, capitalism tamer, or Europe more Christian or less Muslim. The splinters could momentarily be bound together and driven like a stake through the heart of the fiend "Europe" only because the citizens of France lacked the motive and will to be European.

To awaken, unify, and act, Europe needs what the philosopher Peter Sloterdijk has called a mytho-motive, that is, a rich, ambitious imagery and symbolism by means of which its citizens can find their identity and give Europe, as Europe, a mission. Historically what has Europe been as myth and motive? The answer, with all its disturbing resonance, is that whenever Europe has projected itself as Europe it has done so as Empire. Post–Cold War Europeans have to imagine themselves accordingly. Political identity and political project cannot be invented whole cloth, nor can they be adequately sustained solely by loyalty to constitutional principles and procedures. They must arise as the rewriting of deep-seated myths and narratives, a challenge all the more daunting for Europe since the last projections of its Empire myth led to catastrophe. The new Europe must, in Sloterdijk's words, "become the workshop of a metamorphosis of Empire." The rewriting must "replace the principle of Empire itself with that of the union of States in an act creative of a political form that will inscribe itself in the history of the world."[25]

The difficulty of constructing Europe involves not only institutions but also symbols and myths. The French shrunk from greatness with the No vote because their leaders made the arguments for Europe look

---

[25]Peter Sloterdijk, *Si l'Europe s'éveille: Réflexions sur le programme d'une puissance mondiale à la fin de l'ère de son absence politique,* trans. Olivier Mannoni (Paris: Mille et Une Nuit, 2003), pp. 70, 74–75. (My translation from the French translation.)

small, cautious, defensive. The project of European Union is to create an Empire of Rights: democratic, capitalist, expansive, inclusive. So long as Europeans recoil from discovering their animating motive and myth for undertaking this project, the promise of transnational cosmopolitanism will remain rather empty.

That the U.S.-Europe relation is not a fixed polarity is evident in the ever-fluctuating basis of the relation itself. Europeans are in the midst of a great ongoing debate over the relative merits of an "Atlanticist" foreign policy closely tied to the United States as opposed to a "multipolar" vision in which a united Europe would be a counterweight to the United States in world affairs. While the multipolar vision foresees European leadership on behalf of international law and transnational cooperation, it does not require—or actually conform to—Habermas's interpretation of cosmopolitanism. For starters, it is widely recognized that Europe will not achieve diplomatic parity with the United States unless it massively expands its military capacity: the path to a more rule-governed, law-abiding world is a more heavily armed Europe. But would a more muscular Europe actually become less rather than more Atlanticist? More rather than less cosmopolitan? Necessarily more the one because less the other? It is an open question seldom asked. It poses an unexpected variation on Hans Joas's theme: the pressures created by the United States's unacceptably unilateral and dangerous decisions could drive Europe to pursue a "defensive" (post)modernizing military buildup simply to enhance its equity as a dialogue partner with the United States, but such a buildup will inevitably have unanticipated consequences for the European body politic and its presence in the world.

The American orientation is no more settled than the European. Contrary to both Kagan and Habermas, liberal nationalism does not inevitably lead to the Bush doctrine. The Bush doctrine is an extreme interpretation of liberal nationalism in terms of exceptionalism, unilateralism, and global supremacy. Kagan himself had to acknowledge, several months into the Iraq occupation, the drawbacks of putting American policy under the sign of a unilateralist interventionist god:

> It is . . . doubtful . . . whether the American people will continue to support both military actions and the burden of postwar occupations in the face of constant charges of illegitimacy by the United States' closest democratic allies.

Because losing legitimacy with fellow democracies would be debilitating—perhaps even paralyzing—over time, Americans cannot ignore their unipolar predicament. The biggest failure of the Bush administration may be that it was too slow to recognize this truth. . . . The United States can neither appear to be acting, nor in fact act, as if only its self-interest mattered. It must always act in ways that benefit all humanity or, at the very least, the part of humanity that shares its liberal principles.[26]

For someone who swaggered into the international debate a year and a half earlier with large claims about Americans' war-ready fitness for World Power, Kagan's sudden shift from unilateral exuberance to unipolar predicament is a remarkable admission: "Invading Iraq and trying to reconstruct it without the broad benedictions of Europe has not been a particularly happy experience, even if the United States eventually succeeds. It is clear that Americans cannot ignore the question of legitimacy, and it is clear they cannot provide it for themselves."[27] Let us draw the conclusion that Kagan leaves rather muted: the short-sighted haste in making war on Saddam Hussein without broad European cooperation undermined long-term American goals, American interests, American values. In the sobering light of the United States's botched occupation of Iraq, even the Mars-and-Venus difference in the perception of international risks acquires a new meaning: "This gap in perception has driven the United States and Europe apart in the post–Cold War world, and it is difficult to imagine how the United States' crisis of legitimacy could be resolved if this schism persists."[28]

It is not an exaggeration I think to suggest that much of what Kant hoped for when he envisioned a European path toward perpetual peace has already been achieved. European nations have made lasting peace with one another; they have joined with the United States and other democracies around the world to establish many international rules on the use of force; the wealth and influence of Western democracies over the world economy is unrivaled; and their separate and combined military

[26]Robert Kagan, "America's Crisis of Legitimacy," *Foreign Affairs* 83, no. 2 (March/April 2004), p. 85.
[27]Ibid., pp. 72–73.
[28]Ibid., p. 86.

THE ORDEAL OF UNIVERSALISM   165

forces are so superior that no bellicose country could successfully invade any of them, nor even attack with weapons of mass destruction without risking almost certain massive retaliation. The Kantian "federation of free states" already exists through various alliances and treaties, and it draws legitimacy from the United Nations, which it on the whole dominates. This "architecture," as it was frequently called in the Kissinger era, was created at the initiative of the United States. The result, however, has not been perpetual peace. Conflict, insecurity, and war abound. The problems all lie elsewhere: international terrorist networks; failed or weak states that breed civil wars, terrorism, and drug trafficking; rogue states with weapons of mass destruction; and the possibility, indeed the likelihood according to many experts, that weapons of mass destruction will pass into the hands of terrorists willing to use them.

Islam's Geo-Civil War

An uncomfortable truth that often makes dispassionate discussion difficult has to be squarely faced: most of these dangers are coming from the Muslim world. The United States and Europe must learn how to deal with the seemingly endless conflicts generated within and between Muslim countries and between radical Muslims and the West. The extent to which Americans and Europeans work together rather than at cross-purposes will have a lot to do with how bloody and chaotic the coming decades are. American opinion-makers frequently accused Europe after September 11 of feeling so smugly secure from outside threats that it refused to take seriously Washington's more realistic ends and means in the war against terrorism. The reality is quite different. Europeans assess the risks on basis of the fact that Western Europe is home to millions of Muslim origin, many of whom are young, poorly integrated into society, and susceptible to radicalization. While for most Americans the global nature of the crisis in Muslims' relation to the West is condensed in the image of the Twin Towers aflame, Europeans feel it in the very fabric and everyday life of their own societies. European foreign policy toward Muslim countries and Islamic movements always has potentially volatile domestic implications. The United States's utter disregard for this aspect of the European situation exacerbates many global risks and dangers. Worse yet, the Bush administration's refusal to approach Iraq within an international framework, its disdain for the Geneva Conventions, the scandalous treatment of the prisoners at Abu Ghraib and Guantánamo, the collaboration with repressive regimes through "rendition," and American indifference to civilian casualties unnecessarily intensified antagonism especially among young Muslims in Europe, for whom the Palestinians' second intifada had already been propagandized into an emblem of their own estrangement and difficulties within European society. Many began increasingly to heroize the jihadists who flocked to Iraq to oppose the American occupation. In the case of the four suicide bombers in London in July 2005, adoration turned into identification and imitation.

It is sometimes said that Islam is in the midst of a civil war. An apt description in some ways, civil war is also a misleading metaphor. There is something more under way, since the upheavals in the Muslim world are also geopolitical conflicts that do not confine themselves within borders. Islam is embroiled in a *geo-civil war.*

The Islamist terrorist networks epitomize the fusion of geopolitical and civil strife. Al Qaeda is a Saudi-dominated international brigade that recruits among Sunni Muslims around the world; it draws its ideology from Egyptian radicals; it maintains close ties with Pakistani military leaders and security forces; it propped up the Taliban who in turn sheltered it in Afghanistan; and its September 11 attacks were planned as an assault on the United States and at the same time a salvo in the Wahhabi struggle against the Saudi royal family. The Muslim geo-civil war is further amplified by the Israeli-Palestinian conflict and its consequences. In early 2005, the Middle East was a patchwork of foreign occupations, with Israel still occupying Gaza and the West Bank, Syria occupying Lebanon, and the United States Iraq. By the summer of 2006, Israel was caught up in hostilities on two fronts. Having withdrawn from Gaza unilaterally, in effect challenging Palestinians to govern themselves despite the conflicts between Hamas and PLO, Israel found itself under attack from Hamas elements in Gaza and once again sent its troops in. Meanwhile, Hezbollah—whose radical Iran-oriented Islamism is otherwise at odds with the Sunni, Egyptian-oriented Islamists of Hamas—opened the second front by attacking Israel across the Lebanese border; since Hezbollah had ministers in Lebanon's fragile "national unity" government, following Syria's reluctant withdrawal from

Lebanon, Israel treated the Hezbollah intrusion as an act of war by the Lebanese government itself and escalated the conflict with extensive bombings of Beirut and other cities. In fact, as was obvious to all, Hezbollah acted contrary to the Lebanese government; it was fostering civil tension within Lebanon and open conflict with Israel under instructions from Syria and Iran. Syria was eager to deflect international attention from its own involvement in the previous duly elected Lebanese leader's murder, and Iran had at least two motivations: to deflect attention from the international crisis over its nuclear program and, just as important, to heighten Arab animosity toward Israel in order to present Persian Shiism in solidarity with the Arab world. Civil strife and geopolitical conflict endlessly woven and rewoven together.

The Muslim civil war is geopolitical in the sense that the various nations that are roiled by the struggles among Muslims also have complex alliances, rivalries, and conflicts *with one another.* Baathist Iraq and Islamic Iran were bitter, stalemated enemies after their catastrophic war in the 1980s. Iran's only Arab ally had been Syria. Some neoconservatives and hawks touted the idea in 2003 that America might target Syria and Iran after it got rid of Saddam Hussein, but the American action in Iraq instead created the need for the Shiite majority to shore up the fragile new Iraqi government through alliances with, precisely, Syria and Iran. That alliance, in turn, further sharpened the geopolitical divide between Shiites and Sunnis. If Iraq cannot establish political institutions and practices that secure the participation of Shiites, Sunnis, and Kurds, then the Sunni-Shiite divide, which cuts a swath through virtually every country from the Mediterranean to the Indian subcontinent, will deepen and become ever more volatile. Shia are a substantial minority in Pakistan, Kuwait, and Lebanon and are the oppressed majority in strategically crucial Bahrain.[29] Moreover, the violent forms of theocratic extremism that blossom on both these branches of Islam pit extremists against extremists as well as extremists against moderates. The geopolitical conflicts can erupt into religious war; religious conflicts can escalate into civil wars; civil war can overflow borders and become geopolitical.

The potential conflicts within and among Muslim countries are exacerbated by the fact that several nations are ruled by autocratic regimes

---

[29]See Vali Nasr, "When the Shiites Rise," *Foreign Affairs* 85, no. 4 (July/August 2006).

led by a powerful ethnic or religious minority, as Iraq was ruled by its Sunni minority under Saddam Hussein. Saudi Arabia and Syria are key instances. The House of Saud rules Saudi Arabia, and its historic compromise with the fanatical clerics of Wahhabi Islam ultimately nurtured the extremism and resentment that spawned al Qaeda and supplied the insurgency in Iraq with many fighters and suicide bombers. The Assad family and much of Syria's ruling elite are from the Alawi sect, which Shiites acknowledge as Muslim while most Sunnis do not; the Alawi-dominated Baathist party of Syria thus had a history of antagonism with the predominantly Sunni Iraqi Baathists, which further helps to explain why Syria became allied with Iran against Saddam Hussein's Iraq in the 1980s and with Iraq's Shiite-majority government during the American occupation.

Pakistan adds yet another fold in the geo-civil war, for it in effect has two contrary foreign policies in operation at once. President Pervez Musharraf, whom the Bush administration embraced as an enlightened autocrat, balances his alliance with the United States in the pursuit of al Qaeda and the Taliban off against the close ties that his own military and security forces maintain with, precisely, al Qaeda and the Taliban. The Islamist elements within Pakistan are among the most dangerous in the world. Not only do they hold radical aims in solidarity with bin Laden, but they are also firmly entrenched within the state appartatus of a country that has nuclear weapons.

Therein lies another geopolitical crux of American foreign policy: since the two countries whose regimes are most vulnerable to Sunni theocratic movements are Saudi Arabia with its oil and Pakistan with its nuclear weapons, the United States can scarcely afford not to support their autocratic rulers. The neoconservative hawks felt the dilemmas from the beginning: the perpetrators of September 11 were Saudis, but the regime whose oppressiveness fueled their yet more oppressive radicalism also, more literally, fuels the industrial world; and, second dilemma, the task after September 11 of mounting an attack on al Qaeda and the Taliban in Afghanistan could only be achieved by enticing Musharraf, whose government helped create and still supported the Taliban, to assist in their overthrow. The unpleasant truths of Realpolitik that these dilemmas imposed on the administration bizarrely reinforced the neoconservatives' belief that the long-range liberalization and pacification of the Middle East could best be accomplished by destabilizing the region altogether through an invasion of Iraq.

Even this elementary sketch of Islam's geo-civil war highlights how implausible the neoconservatives' domino theory was from the outset. Bush's neoconservative braintrust were beguiled by the opposition of tyranny and democracy, believing that any long-oppressed people would share this Manichaean vision and see their own political choice simplified to democracy-versus-tyranny. Going back to 2001, the Bush administration's geopolitical outlook inverted the priorities in the Middle East in an almost surreal fashion: they dismissed the idea that a settlement of the Israeli-Palestinian conflict would be a decisive first step in undermining Islamist radicalism and concluded instead that overthrowing Saddam Hussein, destabilizing the entire Middle East, and trusting a spontaneous democratic revolution to ultimately put Iraq in the control of pro-Western leaders would be a more direct route to weakening terrorism and resolving the Israeli-Palestinian conflict than . . . resolving the Israeli-Palestinian conflict! They then waited for Yassir Arafat's death and Palestinian elections to open new paths of negotiation, only to see anti-Semitic terrorist Islamist Hamas win the post-Arafat election.

European policy, along with the bulk of antiwar opinion on both sides of the Atlantic, has also been inadequate to Islam's geo-civil war. What the neoconservatives had right was something that American liberals and Europeans still have a hard time admitting, namely, that neither containment nor the status quo was, or is, a viable option in relation to Islamic radicalism. Even before September 11 the so-called status quo was a dynamic, threatening, deteriorating situation. No effective alternative to Bush administration policies will emerge until Democrats and Europeans take this fact as their starting point. By the same token, the inaugural vision that the neoconservatives persuaded themselves of—namely, that the Arab masses in Iraq would embrace democracy by following a pro-Western elite if America simply amputated their oppressive ruler's iron fist—proved so wrong that the administration's own vision of the path to a more liberal, democratic, and peaceful Muslim world oscillates between farce and nightmare, littering the road to democracy with mangled steel and flesh.

Global Neoliberal Religious Conservatism?

The sort of messianic purpose George W. Bush brought to a policy of fighting terrorism, overturning tyranny, and spreading democracy did not manage to rid the policy of its incoherence. The incoherence was there from the beginning in at least three respects. First, the aim of spreading democracy was accompanied by an indifference to "nation building." Second, the simplistic duality tyranny-or-democracy neglected the dangers of disorder in the wake of a tyrant's fall and, even more, the difficulties in inaugurating democratic institutions and practices. And, third, the war against terrorism was wrong-headedly, and dishonestly, associated with Iraqi tyranny, with the result that the American invasion and occupation drew Islamist terrorism *to* Iraq and hastened its spread from Lisbon to Bali, from Sharm el Sheik to London. Only the tenacity and savvy of Iraqi Shia and Kurds saved the policy from immediate ruin.

The American public's embrace of the unilateralist-messianic mission did not sustain itself in face of the costs and casualties of the Iraq involvement. Bush may have originally turned to this rhetoric primarily to manipulate public opinion or to stoke the loyalty of his Christian evangelical "base" or to give his own conscience a center of gravity in the midst of decisions he was ill-equipped to make. Whatever the motives, the rhetoric itself poses two great dangers. It first of all undermines the prospect of close collaboration with our European allies, a collaboration that is necessary not only for what Kagan calls the "legitimacy" of American actions but also for all the substantive tasks of preventing terrorist attacks and fostering liberal democratic state-building in troubled countries. Second, the discrepancy between messianic words and military realties ultimately saps the body politic's capacity and willingness to make lucid, difficult, consequential decisions when necessary.

There are many who utterly disagree with such an assessment and instead anticipate, whether in hope or in fear, that American policy will continue to be guided by messianic motives. Such an outcome is certainly possible, but if it happens it will be the result of political processes that are still gestating and undecided. A messianic foreign policy fits well with the ambition that Bush's key political adviser Karl Rove devised for the Republican Party. Bush-inspired Republicanism yokes together the global assertion of American power, deregulated capitalism, and religious social conservatism. The intellectual design that might join these three elements remains relatively inchoate. In itself this is not a fatal flaw but merely the sign of a still unsettled political process. The electoral strategies devised by Rove have the goal of creating a permanent Republican majority, on the model of the Democratic bloc that held sway in American politics from the New Deal until the 1980s, and as with any major electoral bloc the Republican majority's policies and ideology must be elastic enough not to alienate any of the bloc's vital and heterogeneous components.

The 2004 election proved that messianic nationalism and social conservatism could indeed create a clear majority, as traditional Republicans and independents convinced of Bush's leadership in the "war on terror" were joined by a record number of evangelicals in supporting his reelection. Immediately after the election, however, the president's agenda of deregulated capitalism had little support, as the cold reception to Social Security "reform" showed, and the radical social conservatism of the "culture of life" set off a significant backlash. The permanent Republican majority has yet to solidify, as the 2006 midterm elections showed, and Republicans will continue to experiment with ideology and policy.

Meanwhile, many intellectuals and opinion-makers are hard at work to supply the missing intellectual design. How to make global American power, deregulated capitalism, and religious social conservatism reinforce and ratify one another? One of the boldest efforts in this direction is by the foreign policy expert Walter Russell Mead in *Power, Terror, Peace, and War: America's Grand Strategy in a World at Risk.* Sensing the Bush administration's lack of intellectual coherence, Mead at the same time seems to believe that its actions are in keeping with a kind of abiding historical necessity that has not yet been adequately articulated. In the manner of many authors of foreign policy discourse, he does not advocate so much as analyze and predict. The combination of analysis and prediction slowly acquires a prophetic tenor. What Mead prophesies can be called *global neoliberal religious conservatism.*

His own vocabulary is more colorful: *millennial capitalism* and *American Revival Wilsonianism.* Mead considers the phase in the development of capitalism inaugurated by the New Deal to be definitively over. The old "Fordist era" is being replaced by *millennial capitalism:* "In Fordist capitalism, the market was seen as a dangerous force that had to be harnassed and restricted. Ordinary people needed to be protected from its vagaries. In millennial capitalism, the role of regulation is to protect the existence and efficiency of markets in order to allow wider access to their benefits."[30] As regards foreign policy, Mead regards the upsurge of Christian evangelical themes as a fundamental reorientation, a kind of *American Revival,* in which the internationalism that Woodrow Wilson introduced nine decades ago when he linked democratic values and global involvement is now being radicalized by a disdain for international institutions and fueled by fundamentalist Christianity rather than Wilsonianism's mainline Protestantism which liberals gradually secularized. The result Mead calls *Revival Wilsonianism:* "Returning to Wilsonianism's nineteenth-century roots among missionaries and fervent Protestants, Wilsonian Revivalists are building a strong coalition that binds the Christian right to an assertive, long-term strategy of intervention and, yes, nation-building abroad, even as they embrace a program of strengthening religious values and institutions at home."[31] Thus, the third element in the projected ideology of a permanent Republican majority, religious social conservatism, joins ranks with deregulated capitalism (*millennial capitalism*) and global American power (*Revival Wilsonianism*).

In seeking to fuse these three elements together in a single intellectual pattern, Mead takes up the aspiration of uniting domestic and foreign policy and, in a more rigorous manner than the neoconservatives, of deriving a specifically American foreign policy from a distinctively American democracy. He does so via two ideas.

First, to make the evangelical religious revival in American society and politics compatible with *millennial capitalism,* he evokes a kind of onto-theology of capitalism in which the famous invisible hand partakes of divinity: "I do not believe that the genie of millennial capitalism can be forced back into the Fordist bottle. The quest for greater efficiency, productivity, and dynamism is not a feature of capitalism that can be dispensed with; it is the essence of capitalism, not an excrescence,

---

[30]Walter Russell Mead, *Power, Terror, Peace, and War: America's Grand Strategy in a World at Risk* (New York: Alfred A. Knopf, 2004), pp. 73–74.

[31]Ibid., p . 91.

and come what may it will find ways to fulfill itself. That quest corresponds to the desire implanted in every human individual and perhaps in every living thing to live, to grow, to explore, to search for light, and to fulfill the nature and hidden purpose within."[32] Here Mead obviously slides from his analytic-predictive stance into a profession of faith. An up-to-date faith it is. Christian millenarians of old dreamed of paradise on Earth and found their sacred image of the earthly paradise in the ethic of the Sermon on the Mount. The millennial capitalists evoked by Mead have come up with the idea that God, a few millennia after revealing Himself to a hierarchical tribal society, has changed His earthly affiliations and now devotes His invisible hand to free markets, personal retirement accounts, privately owned utilities, and perhaps—why not?—international security companies as well. So, too, the once pacifist, egalitarian, antimaterialist Jesus has apparently shifted allegiances, substituting Haliburton and Triple Canopy for the meek and the poor.

Second, to make evangelism compatible with realistic foreign policy aims, especially regarding the Islamic world, Mead addresses the obvious concern with the "clash of civilizations." He advances a bold prognosis in which Revivalist America and conservative Islam ultimately become compatible, even cooperative and collaborative. Americans had to learn to distinguish social democracy from communist totalitarianism during the Cold War, and now, Mead argues, they will learn to distinguish an acceptable Islamic conservatism from fascistic fanaticism. House of Saud, sí. Bin Laden, no. Of course Islamic conservatives will impose some restrictions, especially on women, that Westerners may find reprehensible, but tolerance will prevail! Mead predicts that the eventual rapprochement of American evangelicalism and Arab Islam—and one might add conservative Catholicism, solidified since the election of Benedict XVI—will build on their shared reaction against secularism, sexual freedom, and abortion rights. Mead finds precedent for such a "conservative ecumenism" in "the close cooperation and fellowship that can be seen among conservative Catholics and Protestants in the right-to-life movement and elsewhere."[33] The prophecy of global neoliberal religious conservatism is thus complete: neoliberalism as regards

[32]Ibid., p. 196.
[33]Ibid., p. 180.

the global economy; religious social conservatism as regards gender, sexuality, and family life; American military-diplomatic supremacy.

The Achilles heel of Mead's vision is the "conservative ecumenism" among evangelical, Catholic, and Muslim conservatives that he expects to cement global peace. He overlooks how all of these activist conservative religions thrive on prosletyzing. Evangelicals and Catholics in the United States have indeed been able to form coalitions on single issues, but their ability to do so is owed to the very secularization of the political realm that they so often denigrate. Citizens of different faiths find it easy to form coalitions against abortion rights within political parties or outside them in various associations precisely because American politics (thus far) separates religious and political affiliation. For the same reason, conservative Catholics and evangelicals can at the same time differ vehemently over, for example, capital punishment without jeopardizing their shared opposition to abortion rights. In countries that do not effectively anchor their constitution in secularism and pluralism, religious rivalry easily becomes the primary form of political conflict. When the American occupation failed to provide basic security in Iraq, the fear of death drove Iraqis into the arms of their tribal and religious groups. And when ethnic and religious affiliation became the dominant principle in forming political parties, sectarian conflict pushed the country toward civil war.

Ecumenism requires more than overlapping views on the state's role in regulating morality. As it emerged in the era of Vatican II, the ecumenism inspired by the thought of liberal theologians of various faiths amounted to a search for a sense of universalism that could be accepted by all faiths, or at least the Catholic, Protestant, and Jewish faiths then in dialogue. It was acknowledged, more or less explicitly, more or less implicitly, that such a universalism depends on the view that the different religious faiths share a single essential experience of the divine which they elaborate in their own unique manner through their respective symbols and rituals, practices and doctrines. Competing beliefs were held to be complementary beliefs. Such a relativizing of the forms of worship, such a loosening of each faith's doctrine's claim to dogmatic truth, has certainly lost ground in the intervening decades. It has proved very difficult for believers to embrace a pluralistic attitude toward belief. That leaves tolerance. Religions can tolerate one another's existence even though they cannot grant one another the same share of the truth. However, religious communities are most likely to coexist in mutual tolerance if they maintain what Jefferson called the "wall of separation" between religion and state; full-fledged religious tolerance requires pluralism in the *political* sphere though not within religion itself. When the wall of separation is breached, or never built in the first place, tolerance becomes precarious. If religious affiliation begins to coincide with political allegiance, if at the extreme, as in Iraq, political parties are formed on the basis of religious affiliation, sectarian conflict becomes ever more probable.

In many societies, the transformation of religion into politics is on the rise. In Africa, the prosletyzing fervor of evangelical Christians is changing the entire landscape of political conflict. The journalist Eliza Griswold makes the observation that the 10th parallel north of the equator is "a kind of religious faultline around the world," a demarcation between Buddhism and Islam in southern Thailand and between Christians and Muslims in Africa and Indonesia. There is great reluctance to ponder the consequences.

> For centuries, this line has been largely inconsequential [across Sub-Saharan Africa], buffered by animism, but today, due to the contentious presence of mostly evangelical Christians on the other side of that line, some theologians think we're looking at the future of conflict in Africa. In Africa, for example, since 1900, the number of Christians has grown from approximately 9 million to 330 million. In Northern Nigeria, in particular, according to the Nigerian government, this Christian-Muslim conflict has cost 50,000 lives in a three-year period, yet we know nothing about that here.[34]

The sectarian religious nature of the civil strife in Côte d'Ivoire has also gone largely unreported. According to Andrew Rice, writing in *The New Republic*, "Laurent Gbagbo, a Christian politician from the country's south, won election to the presidency in 2000. Rebel movements popped up in the Muslim north of the country, led by men who disdained

---

[34]Remarks made at a roundtable on "Faith at War: Reports from the Islamic World," sponsored by the Pew Forum on Religion and Public Life and the Council on Foreign Relations, May 4, 2005. Transcript available at http://pewforum.org/events/print.php?EventID=78.

Gbagbo's open faith—Gbagbo surrounded himself with evangelical preachers—and his thuggishness, and a bloody civil war broke out."[35] Zambia's president has declared his country a "Christian nation." Muslim efforts to introduce *sharia* have intensified conflicts in Kenya as well as Nigeria. And in Uganda, Islamic fundamentalists are ripe for renewed rebellion against the expansionist Pentecostals and evangelicals who already boast as many as twenty-five million adherents, including the country's enthusiastically born-again first lady.

The symbiosis between the American Revival and African evangelization is, in short, encouraging violent conflict rather than new forms of ecumenism. Meanwhile, Arab monarchs and oligarchs are contributing millions to spread Islam, often a radicalized Islam, in the same countries where American churches and missionaries devote huge resources to converting Africans to Christianity. African born-agains, who watch American as well as African televanglists, are highly supportive of American global power and Bush's policies, just as the Revival Wilsonians want. But, as Andrew Rice discovered in covering one of Uganda's rising evangelical preachers, "Increasingly, Africans like [Pastor Martin] Ssempa support the United States not only because of their ties to American evangelicals or because they believe Washington is fighting terrorism and promoting democracy, but for a baser reason: the United States is killing Muslims."[36] The potential *political* resemblance between evangelicalism and Islamism lies less in their shared conservatism regarding family and sexuality than in their tendency to form tightly knit groups and movements around charismatic leaders. The ambitious preacher, unlike the Catholic priest, is relatively unhampered by institutional controls and constraints. He is free to exert charismatic leadership over whatever body of followers he can mobilize. This is why the Protestant demagogue can aspire to personal wealth and political power *via* his "ministry." The Sunni jihadist leader can evoke religious authorities to support the most violent actions against infidels, Jews and crusaders, and apostates. And every aspiring ayatollah can issue whatever fatwa he chooses to enhance his power and direct his followers. All these entrepreneurs of the soul are threatening to alter the very nature of politics.

[35]Andrew Rice, "Evangelicals v. Muslims in Africa," *New Republic,* August 9, 2004.
[36]Ibid.

Religious or ethnic sectarianism is often the swiftest means to political power in failed states or in a power vacuum created by the sudden collapse or, as in Iraq, overthrow of a tyrannical regime. Wherever such sectarianism arises, the prospect of democracy fades; democracy is promoted only by enhancing the values and institutions of pluralism and secularism. As Iraq tries democracy, the outcome has come to depend, at best, on some kind of uneasy balance among ethnically and religiously defined political forces; the seeds of civil war were firmly planted in the soil of the country's constitution and political parties. How much worse the prospects of democracy are in countries where political clashes are likely to take the form of confrontations between Muslims and Christians. Republicans are playing a dangerous game by trying to fuse evangelical Christianity and foreign policy, even if it is predominantly only in the minds of evangelical Christians themselves. The evangelicals will continue to expect support and legitimacy for their international prosletyzing, which can only undermine the professed goal of spreading democracy. Worse yet, if a clash of civilizations by proxy is in the offing, the outcome could easily prove far more destructive and deadly than the proxy wars that made the Cold War anything but cold for millions in Southeast Asia, Africa, and Latin America.

In politics it is far more dangerous to strike a bargain with God than with the devil.

No Exit

President Bush made *freedom* the universalist battle cry of American foreign policy. His words inspired many by awakening the hope that the United States had turned a corner after decades of Cold War Realpolitik and would now use its immense wealth and power simply to liberate people from repressive regimes. In Poland and Hungary, the war in Iraq seemed a long-overdue resolve to attack totalitarianism in whatever guise. In France and Germany, by contrast, Bush's unilateral saber-rattling stirred anxieties and contempt, and war in Iraq looked unnecessary, ill-advised, illegal. The invasion of Iraq also provoked simple skepticism: What did it have to do with terrorism? Were American aims and means realistic? The answer to the first question was, *Nothing*. The answer to the second turned out to be, *No*. The unprecedented task that had to be faced after September 11 was to break up al Qaeda, establish international cooperation in policing Islamist networks, and mount an educational, diplomatic, and ideological offensive to counteract the radicalization of Muslim, especially Arab, youth. Bush chose war in Iraq instead.

Half a decade after September 11, the balance sheet is grim: American troops are overextended; the United States can neither end nor abandon the civil war in Iraq; al Qaeda survives and has metamorphosed into loosely connected, supple networks around the world; American actions in Iraq and Guantánamo have increased the recruits and sympathizers of Islamic radicalism; American allies doubt more than ever the United States's motives *and* its ability to achieve what it claims.

What, then, will now be the fate of the idea of freedom that was summoned to justify the war in Iraq? The idea itself was expressed by Bush in a rhetoric at once absolute and vague, as though announcing a new liberation theology. It often merely sounded like a rehash of the typical rationalization for the American use of force, but it also seemed a precipitous venture onto a new, perhaps necessary terrain for American foreign policy. To be sure, American presidents have never been able to persuade the citizenry to war unless they did so in the name of freedom, claiming to protect freedom at home and extend freedom abroad. Operation Iraqi Freedom exemplified the pattern and enjoyed a simplifying clarity of meaning. *Freedom* meant *liberation from a tyrant.* As soon as the tyrant fell, however, so did the scales from the daydreamers' eyes. Democracy did not spring full-grown from the ruins of tyranny, and Iraqis did not spontaneously and gratefully erect a pro-Western state.

The freedom-versus-tyranny opposition had overlooked the ordeal of universalism that is at the heart of democratic creativity. And among the many other setbacks and losses suffered in Iraq, the Bush administration may also have squandered the very idea that freedom and democracy should be guideposts of foreign policy. The idea has lost persuasiveness, indeed credibility, even plausibility. Neither allies nor voters nor global public opinion believes the Bush doctrine capable of stopping terrorism or of spreading democracy. By failing to secure Iraq, the American occupation ruined the chance to furnish Iraq a genuine tutelage in democratic inauguration and undermined the conditions by which Iraqis might create the foundations of a democratic state. Ill-suited for nation building and democracy promotion in the first place, the administration failed to prepare for the postwar situation. Their ineptness was exacerbated by their disdain for the sort of international or multilateral effort that would have been needed to forge a peaceful transition to democracy.

Under pressure to show progress in the midst of the violent insurgency, the administration consistently turned to the quick-fix accoutrements of democracy while continuing to neglect the institutions and practices that might ultimately sustain actual freedoms. Sovereignty, elections, and constitution were hurried into place. Majority rule became the one principle on which to claim that Iraq was becoming democratic. Leaders of Iraq's Shia majority did not fail to grasp that their embrace of this principle was the quickest and surest route to power. As the deadlines for a constitution neared, the Bush administration threw its support behind compromises that greatly expanded the role of Islam in making and interpreting Iraqi laws and granted considerable autonomy to the oil-rich Shiite South and the oil-rich Kurdish North, thereby seriously undermining secular democracy and women's rights as well as Sunni participation in the political process. Bowing to the conservative leaders of a religious majority might look to some like a first step toward fulfilling Walter Russell Mead's prophecy of global conservative ecumenism, but in reality it heightened the prospect of sectarian strife and civil conflict. Forgetful of America's own founding fathers, who were preoccupied with crafting a constitution that would discourage the formation of *factions* and protect against the *tyranny of the majority*, the Bush administration gave its blessing to an Iraqi constitution that encourages factions and legalizes the tyranny of the majority.

A value long associated with American habits and practices, namely, the value of the individual, has been disregarded in the Bush administration's effort at nation building. Democratic freedoms, rights, and participation hinge on individuals *as* individuals. Rights are individual, freedoms are individual, political participation is individual. Of course movements and parties are collective undertakings energized by trans-individual identifications and emotions, organizations and symbols, and power itself is action *in concert,* to evoke once again Hannah Arendt's formulation. Democracy's power, though, lies in the action in concert of *individuals,* individuals endowed with specific rights and freedoms, including the right and freedom to dissent from the action in concert. It is such individual right, freedom, and participation that the United States so dramatically failed to nurture in Iraq. In the absence of civil order, militias arose in the anarchic fear of death; political factions formed along sectarian ethnic and religious lines; parties exacted blind loyalty rather than fostering open debate; the majority licensed itself to tyrannize individuals and minorities.

In the vacuum created by the fall of Saddam Hussein, Iraq was caught between democratic striving and sectarian mobilization. Two episodes capture the antagonism. The individual exercise of rights, freedoms, and participation found its embodiment and emblem in the men and women holding up their ink-stained fingers in Iraq's first national election. When, by contrast, the constitution was being decided, sectarian mobilization and manipulation were the order of the day, as political parties used their militias and gun battles to affect the deliberations and issued "orders" telling their followers how to vote in the constitutional referendum.

The quagmire in Iraq will unfortunately erode most Americans' commitment to fostering democracy and struggles against tyranny. Isolationism remains an easily tread path in American political consciousness, even as the international state of affairs makes such a reflex utterly dangerous and senseless. Among Democrats as well as antiwar activists, the view that the war was a mistake has not translated into a forceful understanding of what *should* be done. Yet there is no escaping the need to stabilize Iraq, produce a new and more lucid vision of the war against terrorism, and push liberal and democratic reform in the Muslim world. The leviathan cannot now retreat to calmer waters any more than it can continue to pretend that by "taking the war to the enemy" it has kept violence and death at bay. The failures of the Bush administration have only made these tasks of global involvement more urgent while leaving the diplomatic, political, and military means of accomplishing them in disrepair. Islam's geo-civil war is now an aspect of universalism in the simply empirical sense, that is, of universal history as the raw fact that the effects of events and acts are increasingly felt worldwide. Freedom, cosmopolitanism, and ecumenism are potent universal ideals, but none of them is in the hands of historical necessity. They are in the hands of historical actors, whose geo-imagination must rise to the new global realities. Freedom does not spread out across the globe because the masses are awestruck by the shock of American power. Cosmopolitanism does not advance by adhering to the utopian expectation of a world of law-abiders devoid of passions, interests, and irrational identifications. And ecumenism has no chance so long as the world's believers adhere to their own beliefs as absolute truth and continue to seek power and territory in the name of their God.

Unity of mankind means: No escape for anyone anywhere.
—MILAN KUNDERA

CONCLUSION: PRELUDE
TO THE UNKNOWN

Ideas and Errors

The attacks of September 11 revealed the global reach of Islam's geo-civil war. The Bush administration based its response on two quite valid assumptions: first, the unstable and deteriorating situation in the Muslim world, especially the Middle East, made deterrence or containment designed to maintain a status quo untenable; and, second, existing international laws and institutions are inadequate to the challenges posed by tyrannical regimes, failed states, and governments harboring or supporting terrorists. The policy that was built on these assumptions has largely been a dangerous failure. The administration's errors have been grave. It gave unfounded priority to overthrowing Saddam Hussein when in fact such an adventure distracted from the war against terrorism, and it undertook the war in Iraq in a manner that mocked international law and institutions rather than trying to revise and strengthen them through a precedent-setting multilateral action. It then so badly misconceived the occupation that Iraq, whose reconstruction as a thriving modern democracy was the ultimate justification for the intervention, was turned into the newest theater of global terrorism and Islamic radicalism. Nevertheless, the errors and failings do not refute the two inaugural assumptions. Most critics of administration policy unfortunately resist or reject those assumptions, and Bush's political opponents have consequently stumbled in answering the two most urgent questions: how to address the still unstable, now more rapidly deteriorating state of Islam's geo-civil war, and how to craft effective laws and institutions to deal with states that fail in their sovereign duty to protect their citizens.

The more severe the situation in Iraq became, the more American foreign policy debate seemed caught between "Stay the course," even though the course had proved ineffectual and harmful, and "Bring the troops home," when any such isolationist posture in the midst of an inescapable global conflict is dangerous and self-defeating. Desperately needed is a debate that manages to eschew both stances and all the half-measures that are merely a compromise between them. That task is made more difficult by the fact that isolationist illusions have a strong hold on American attitudes, while on the other hand Bush's messianic rhetoric has severely undermined the credibility of the United States's commitment to advancing freedom and democracy.

The idea of freedom and democracy has been degraded by the coarse rhetoric and simplistic concepts used to justify the war in Iraq, just as the reality of freedom and democracy was undermined by the expedient compromises and willful falsehoods regarding Iraq's reconstruction and new constitution. The current morass encourages intellectual and political flight. The antidote that political thought must provide is to measure political ideas anew against the errors in Iraq. The animating ideas of democratic thought will prove of little worth unless they are now sharpened and deepened in response to all that has gone wrong. An unvarnished appraisal of the torment of Iraqi state building must be brought to bear on the fundamental questions regarding freedom and democracy in American foreign policy, specifically, the freedom-versus-tyranny theme, the relation between freedom and self-rule, and the contrary pull of democratic striving and sectarian mobilization.

Arendt with Berlin

The great theme of freedom-versus-tyranny misled Bush and the neoconservatives. For while tyranny crushes freedom, crushing a tyrant does not necessarily give rise to freedom. Freedom-versus-tyranny neglects the relation between *freedom* and *self-rule,* which is ultimately the very relation of freedom and democracy. In the wake of the American and French revolutions, freedom and self-rule became yoked together in political experience, as did freedom and equality. But just as there are permanent conflicts between freedom and equality, so too freedom and self rule have an uneasy relation. Isaiah Berlin highlighted the unease in his classic essay "Two Concepts of Liberty" by distinguishing negative liberty (the freedom *from* . . .) and positive liberty (the freedom *to* . . .). Negative freedom is in essence "simply the area within which a man can act unobstructed by others," so that defining and securing such an area comes down to drawing the lines that limit the power of the state with regard to "a certain minimum area of personal freedom that must on no account be violated; for if it is overstepped, the individual will find himself in an area too narrow for even that minimum development of his natural faculties which alone makes it possible to pursue, and even to conceive, the various ends which men hold good or right or sacred. It follows that a frontier must be drawn between the area of private life and public authority."[1] Negative freedom thus defined is a value completely distinct from the value of self-rule. Indeed, it does not require self-rule: "liberty in this sense is not incompatible with some kinds of autocracy, or at any rate with the absence of self-government. . . . It is perfectly conceivable that a liberal-minded despot would allow his subjects a large measure of personal freedom. . . . Self-government may, on the whole, provide a better guarantee of the preservation of civil liberties than other régimes, and has been defended as such by libertarians. But there is no necessary connexion between individual liberty and democratic rule."[2]

Writing with the Cold War and totalitarianism in mind, Berlin associated positive liberty with revolutionary schemes that saw humanity under a kind of historical compulsion to realize some "true self." Positive liberty "derives from the wish on the part of the individual to be his own master," and it developed historically so as to have come ultimately into conflict with negative liberty understood as "the freedom that consists in not being prevented from choosing as I do by other men."[3] Self-mastery gave rise to the sense of freeing oneself, for example, from enslavement to the passions, as though one were two selves. The dominant self is the real, ideal, higher, or autonomous self in contrast to the

[1] Isaiah Berlin, "Two Concepts of Liberty," in *Four Essays on Liberty* (New York: Oxford University Press, 1969), pp. 122, 124.

[2] Ibid., pp. 129–30.

[3] Ibid., p. 131.

merely empirical lower, irrational, passion-driven, impulsive self, which "is brought to heel." According to Berlin, when this conception of freedom as self-mastery and self-realization gets taken up within political thought it mushrooms into revolutionary ideologies that prepare the way for the worst oppressions:

> the real self may be conceived as something wider than the individual (as the term is normally understood), as a social "whole" of which the individual is an element or aspect: a tribe, a race, a church, a state, the great society of the living and the dead and the unborn. This entity is then identified as being the "true" self which, by imposing its collective, or "organic," single will upon its "members," achieves its own, and therefore their, "higher" freedom. . . . Once I take this view, I am in a position to ignore the actual wishes of men or societies, to bully, oppress, torture them in the name, and on behalf, of their "real" selves.[4]

Berlin's criticism of positive liberty penetrates the conceptual core that the totalitarian practices of the twentieth-century took from the tradition of revolutionary thought that runs from Rousseau to Lukàcs, and his reflection on negative liberty expresses an experience of freedom that is indelibly a part of the modern political idea of pluralism. Nevertheless, it is not so clear that he thereby exhausts the possible meanings of positive liberty, nor that he establishes that positive liberty is intrinsically destructive of negative liberty.

One of the great missed dialogues in twentieth-century political thought is a debate between Isaiah Berlin and Hannah Arendt, for Arendt's essay "What Is Freedom?" adduces a very different conception of positive liberty from Berlin's even as she too criticizes totalitarianism, the Rousseauistic true self, and the ideal of sovereignty. It is not a matter of trying to reconcile Berlin and Arendt, since their philosophical and political divergence is real. Rather, a new space of reflection opens from the fact that neither thinker quite anticipates the antagonistic view held by the other regarding positive and negative liberty.

Arendt's supreme value is self-rule, the legacy from Greek democracy that the highest form of human self-realization is participation in a po-

---

[4]Ibid. pp. 132–33.

litical community of equals, that is, those who are equal in ruling and being ruled. Arendt is squarely on the side of positive liberty in Berlin's usage. However, for her there is not a dividing line between self-realization and human plurality. On the contrary, the polis is a space of freedom insofar as it is where the plurality of empirical individuals manifests itself. In her reading of the history of political experiences and ideas, the positive liberty established in political community precedes, historically and logically, every other experience of liberty:

> it seems safe to say that man would know nothing of inner freedom if he had not first experienced a condition of being free as a worldly tangible reality. We first become aware of freedom or its opposite in our intercourse with others, not in the intercourse with ourselves. . . . Freedom needed, in addition to mere liberation [from the necessities of life], the company of other men who were in the same state, and it needed a common public space to meet them—a politically organized world, in other words, into which each of the free men could insert himself by word and deed. . . . Without a politically guaranteed public realm, freedom lacks the worldly space to make its appearance.[5]

Arendt thus identifies positive liberty with participation in the politically guaranteed public realm, and she sees this freedom as the historical and existential precondition of all other experiences of freedom, from the pursuit of individual ends unimpeded by others (negative liberty in Berlin's sense) to the inner freedom of thought, belief, and conscience as the philosophical, religious, or moral dialogue with oneself. Positive liberty so understood is not openly antagonistic to negative liberty; it is its precondition. Even where positive liberty does not exist except as the implicit remembrance of a lost public realm, it conditions the experience of inner freedom. By the same token, Arendt is acutely aware that modern totalitarianism has cast suspicion on politics as such: "The rise of totalitarianism, its claim to having subordinated all spheres of life to the demands of politics and its consistent nonrecognition of civil rights, above all the rights of privacy and the right to freedom from

---

[5]Hannah Arendt, "What Is Freedom?" in *Between Past and Future: Eight Exercises in Political Thought,* (enl. ed.) (New York: Penguin, 1977), pp. 148–49.

politics, makes us doubt not only the coincidence of politics and free-dom but their very compatibility."[6] Berlin stood squarely in such doubt and responded to totalitarianism by asserting the supreme value of lib-erty as the private space where the government does not intrude on the individual's choices; Arendt responded to totalitarianism by asserting the supreme value of the public space where citizens ultimately might participate as equals in ruling and being ruled.

Individual freedom and self-rule do not, therefore, fit smoothly to-gether even though both are essential elements of modern democratic thought. Yet, the sense in which Arendt and Berlin are *not* at odds is also crucial to understanding this rivalry of supreme values. Arendt ties pos-itive liberty and Berlin negative liberty to the twin values of individual-ity and plurality. For Arendt, democratic self-rule entails a citizenry that participates in public affairs *as* individuals with a plurality of perspec-tives, values, opinions, talents, interests, and passions. For Berlin, indi-vidual liberty is inseparable from the plurality of human goals and proj-ects: "The world that we encounter in ordinary experience is one in which we are faced with choices between ends equally ultimate, and claims equally absolute, the realization of some of which must inevita-bly involve the sacrifice of others. Indeed, it is because this is their situ-ation that men place such immense value upon the freedom to choose."[7] Thus, even as negative liberty and self-rule are rival values, both affirm individuality and plurality because neither flourishes unless individual-ity and plurality flourish.

The errors in Iraq give a measure of these rival values. Individuality and plurality were quickly sacrificed as the American failure to secure order in Iraq allowed factions and militias in the guise of political par-ties to take charge of the political realm. The dominance of the militias and religious leaders foreclosed the possible deliberation and debate among *citizens* and replaced it with demands for obedience and loyalty on the part of *followers*. Liberty in Berlin's sense and self-rule in Arendt's must be yoked together in perpetual rivalry for democracy to take hold. Berlin's insight that freedom is distinct from self-rule does not so much support liberal autocracy as remind democrats that freedoms must be

[6]Ibid., p. 149.
[7]Berlin, "Two Concepts of Liberty," p. 168.

protected *from* not merely *by* elections and majority rule. Satisfying it-self with the freedom-versus-tyranny theme, the American occupation neglected essential preconditions of liberty and self-rule: the security that comes from a civic authority with a monopoly on the legitimate use of violence and an inaugural political process that values citizens' individuality and plurality.

Liberty without Democracy versus Democracy without Liberty?

The distinction between liberty and self-rule has returned in foreign policy discussions in debates over support for liberal autocracy as an alternative to democracy promotion. Fareed Zakaria, for example, holds a decidedly cautious view of democracy's global prospects, especially in the Arab world, and argues that to establish liberty, even by autocratic means, is a more urgent and rational priority than trying to foster democracy in countries that are unpracticed in individual freedom and the rule of law. Constitutionally guaranteed basic liberties regarding property, speech, association, and religion—and an independent judiciary capable of protecting those liberties—must *precede* democracy. Zakaria especially questions what he considers the fetish of taking elections to be the measure of a country's progress toward liberal democracy: "Across the globe, democratically elected regimes . . . are routinely ignoring constitutional limits on their power and depriving their citizens of basic rights."[8] He especially privileges the so-called East Asian Model: "capitalism and the rule of law first, and then democracy. South Korea, Taiwan, Thailand, and Malaysia were all governed for decades by military juntas or single-party systems. These regimes liberalized the economy, the legal system, and the rights of worship and travel, and then, decades later, held free elections."[9]

There is one glaring flaw in the idea that troubled nations should be steered toward autocracy: it's not all that easy to find benign and enlightened despots. Although historical circumstances have certainly existed where autocrats prepared the way for democracy, rulers with a taste for autocracy are not by nature inclined to give their citizens a wide area

[8]Fareed Zakaria, *The Future of Freedom: Illiberal Democracy at Home and Abroad* (New York: Norton, 2003), p. 17.
[9]Ibid., p. 55.

of basic freedoms, especially freedom of speech and association. And how are speech and association inhibited except by repressive measures that are utterly incompatible with the rule of law and an independent judiciary? The autocracy thesis has its antecedents in Cold Warriors like Milton Friedman and Jeanne Kirkpatrick, who supported repressive regimes so long as they sustained a market economy; their epigone of enlightened autocracy was General Pinochet.[10]

Policy analysts associated with "democracy promotion" are harshly critical of the "economics first, democracy later" formula. Thomas Carothers turns the tables by pointing to Egypt, where it is the "stagnant, semi-authoritarian political system that has undermined the efforts on economic reform. . . . Lacking popular legitimacy that could come from genuine democratic processes, Mubarak badly needs the economic lever of reward and punishment that Egypt's statist economic structures give him to co-opt opponents and reward supporters." Carothers's description of Egypt does not, it seems to me, decisively refute Zakaria; it merely suggests how an authoritarian political system and a retrograde economy reinforce one another. The question at issue has to do with where such a political and economic knot should first be severed. Another of Carothers's objections brings out a more crucial weakness in the autocracy thesis, namely, that the formula of "deferred democratization" neglects the fact that "global political culture has changed" and "people all around the world have democratic aspirations."[11] Larry Diamond makes a similar point: "It is just not possible in our world of mass participation and democratic consciousness to give people the right to think, speak, publish, demonstrate, and associate peacefully, and not have them use those freedoms to demand as well the right to choose and replace their leaders in free and fair elections."[12]

These criticisms clearly throw the autocracy thesis into question, but they also leave some doubts concerning the assumptions made by the

---

[10]Though Zakaria does not acknowledge these intellectual predecessors of his position, he does—in a strange aside on Russia—suggest that Putin may turn out to be a "good czar" leading Russia on a path more like Chile's than postcommunist Poland's: "The Pinochet model is certainly possible; Pinochet did eventually lead his country to liberal democracy"(!). Ibid., p. 95.

[11]Thomas Carothers, "Zakaria's Complaint," *The National Interest* (Summer 2003).

[12]Larry Diamond, "The Illusion of Liberal Autocracy," *Journal of Democracy* 14, no. 4 (October 2003), p. 169.

democracy advocates themselves. Consider the keywords in their assessment: *popular legitimacy, democratic aspirations,* and *mass participation.* Let us break this triad down differently. For although these three elements have become part of political culture worldwide and are indeed essential features of democracy, they do not inevitably gel. The ordeal of state building in Iraq has painfully revealed that the need for popular legitimacy and the fact of mass participation do not inevitably further democratic aspirations. They can, to the contrary, mobilize sectarianism. Charismatic leaders can mobilize the masses by evoking highly charged religious or ethnic identifications and claim popular legitimacy through electoral or nonelectoral means.

Popular legitimation and mass participation have, since they were invented by the American and especially the French Revolution, taken nondemocratic as well as democratic forms. Modern political culture was bequeathed popular legitimacy and mass participation by those Enlightenment-era revolutions, but totalitarianism also thrives on these twin norms; fascist, communist, and populist rulers throughout the twentieth century mobilized the masses and sought popular legitimacy.[13] The modern democratic revolutions created popular legitimacy and mass participation, but modern democracy itself requires *something else* as well. It has to secure individual rights and freedoms, protect plurality, and institutionalize a separation and balance of powers.

Is democracy promotion then caught in an ever impossible attempt to square liberty and democracy? Is most of the world condemned, at best, to democracy without liberty or liberty without democracy? The controversy between democracy promotion and liberal autocracy seems to end up at such an impasse because each side adheres tenaciously to its own premise, democracy first or liberty first, and has no trouble finding abundant examples to rebut the other's premise. The debate inadvertently reveals the key problem, namely, that democratic inauguration never enjoys stable footing or a guaranteed path. Democracy is ungrounded in the sense that its inauguration and its survival require "the unreliable and only temporary agreement of many wills and inten-

---

[13]Claude Lefort, "The Logic of Totalitarianism," in *The Political Forms of Modern Society: Bureaucracy, Democracy, Totalitarianism,* ed. John B. Thompson (Cambridge: MIT Press, 1986), pp. 273–91.

tions."[14] Moreover, no single idea, including liberty and self-rule, can guide the creation of democracy without coming into conflict with other guiding ideas.

This ultimately brings us back to the ordeal of universalism. Democratic commitments always express themselves in a universal principle, yet no single universal principle unequivocally expresses democratic commitment. This enlivening contradiction of democracy is already to be seen in the motto that inaugurated France's democratic revolution: *liberté égalité fraternité*. I imagine a body politic that overthrows tyranny and declares *Liberté!*, thereby throwing itself off-balance and instantly realizing that if liberty, hitherto a privilege of aristocracy, was no longer to depend on status and tradition, it must be secured by law and extended to every individual, and so declares *Égalité!*, only to realize, instantly, that the liberty of individual pursuits, thus universalized, will generate new and unpredictable inequalities and threaten the very cohesiveness of life in common, and so, righting itself once more, shouts *Fraternité!* But *fraternité* can take many forms: national identity, civic solidarity, chauvinism, cosmopolitan world-citizenship, racism. So, too, *liberté* and *égalité* are open to conflicting interpretations, and more fundamentally, no matter how interpreted, they tend to conflict with one another, which is of course what requires *fraternité* in the first place. The democratic body politic thus always teeters, like a tightrope walker, in that unobjectifiable spot between standing and falling, ever inventing the precarious balance of *liberté égalité fraternité*. Democracies thrive so long as they live perpetually off-balance without falling.

Democratic Striving and Sectarian Mobilization

In so-called revolutionary situations, that is, where an ancien régime falls, the resulting power vacuum has to be filled by an improvisation, by unscripted actions in concert. When democratic forces do not take hold in the interval during which power is being reinvented, nondemocratic forces will. Democratic forces unprepared for the sudden power

---

[14]Hannah Arendt, *The Human Condition* (Chicago: University of Chicago Press, 1958), p. 201.

vacuum are likely to be displaced or crushed by highly disciplined groups willing to use violence. The fate of the Girondists in the French Revolution and the Menchiviks in the Russian Revolution emblematizes the defeat of democratic aspiration. As the foreign power that overthrew the tyrannical rule of Saddam Hussein, the United States's responsibility was to preserve and prolong the interval for the reinvention of power, but it was not up to the task, which first and foremost required securing civil order and normal economic activity. Militias, charismatic religious leaders, warlords, foreign terrorists, and ancien régime insurgents soon stepped into the struggle over what shape the political realm would have. In the race between democratic aspiration and sectarian mobilization, democracy progressively lost ground.

The role of sectarian mobilization was enhanced in two steps. Already in the national elections in January 2005, the use of nationwide slates rather than election by districts discouraged citizens' grassroots involvement in shaping issues and selecting candidates. It fostered instead the mobilization of Shiites as a bloc that could be most readily united by their respect for their religious leader. Second, the U.S.-led negotiations over the constitution then sealed the exclusion of Sunnis from the process—abetted of course by their own virtual boycott of the January vote—and turned constitution making into a compromise between Kurds seeking as much regional autonomy as possible and Shiites seeking to extend their majority rule as far as possible. The constitution creates in effect three distinct zones, each of which reflected a completely different political reality: the Kurdish zone where the relatively autonomous parliament and police, along with the peshmerga militias, had already established a relatively stable state (though death squads operated in the name of antiterrorism and with U.S. acquiescence all during the constitutional negotiations); the Shiite South, where the daily, often violent struggle between competing groups was shaping some ultimate compromise between Iran-inspired radicals and the traditional quietistic clergy; Baghdad, where insecurity reigned, and the predominantly Sunni area, where international jihadists and Baathist insurgents held sway.[15]

[15]Critics of the American reluctance to change strategy and tactics for securing Iraq against the insurgency have especially focused on Baghdad. See, for example, Andrew F. Krepinevich, Jr., "How to Win in Iraq," *Foreign Affairs* 84, no. 5 (September/October 2005), pp. 87–104.

It has been argued that a division of Iraq into largely autonomous, loosely federated Kurdish, Shiite, and Sunni zones was all but inevitable in light of the history of the Sunni minority's dominance under Saddam Hussein, the considerable autonomy already enjoyed by the Kurds, and the Shiites' religiosity and close ties with Iran. As Peter W. Gailbraith succinctly put it on the eve of the constitutional referendum, "Iraq's Kurds don't want to live in pluralistic, multiethnic, centrally governed Iraq, and they don't have to. . . . The Shiites do not want to live in a secular society." The constitution reflected these positions, but it was less a constitution than a peace treaty worked out between Kurds and Shiites, granting each of them the means to pursue their autonomy and at the same time hold off a resurgence of Sunni dominance. A series of deals brokered by the American ambassador, the constitution was not truly authored by those who are pledged to live under it. When Kurds were faced with the American acceptance of putting clerics on the Supreme Court, they decided to forgo their opposition to the Islamicization of the judiciary and instead succeeded in "stripp[ing] the Iraqi Supreme Court of jurisdiction over Kurdistan's laws"! While Galbraith argues that the constitution was the only plausible outcome for avoiding the complete breakup of Iraq, with all the civil strife and regional conflict that would entail, he also pinpoints what I consider the constitution's fatal flaw, namely, that Baghdad itself cannot possibly be governed on principles of ethnic autonomy and religious separatism since its population is thoroughly diverse. As the constitution was being finalized, the Iraqi capital remained "the center of a dirty war between Sunnis and Shiites" and had a murder rate "exceed[ing] one thousand per month, not including the dead from car bombs, and many of these are victims of sectarian conflict."[16] It is difficult to see how Baghdad will not remain the site of continuing civil war, a bloodier Belfast with an endless supply of animosities and extremists fed into it by two "communities" that let themselves be defined by religious schism.

Was a liberal democracy, as many opponents of the war argued from the beginning, impossible in Iraq all along? The reasons usually given were the Iraqis' lack of democratic experience and susceptibility to

[16]Peter W. Gailbraith, "Last Chance for Iraq," *New York Review of Books* 52, no. 15 (October 6, 2005).

sectarian division. But these do not inevitably amount to insurmountable obstacles to democracy. They are, however, the very obstacles that the Bush administration so tragically discounted. The outlines of what might have been required to permit Iraq a transition from dictatorship to democracy have emerged from the various prewar plans that the administration rejected and the assessments that experts have made in hindsight: half a million troops on the ground for several years; an occupation force trained to secure the country, disband militias, and retrain and reindoctrinate the Iraqi army; a multilateral or UN-led administration, not under American control, charged with refurbishing the economy and overseeing democratic participation, balancing the double imperative of putting power in the hands of Iraqi as expeditiously as possible and building democratic habits from the local level up. This alternative would have been a daunting and expensive commitment, which the public and American allies might well have rejected. One suspects that the architects of Bush's policy chose their high-risk path not only because they wanted to demonstrate American unilateralism for all the world to see, but also because the alternative was much harder to sell politically and diplomatically. Their actions have ended up making such political and diplomatic persuasion even more difficult in the future, even as events have made the task more urgent.

If Islam's geo-civil war could simply be contained, that is, if its harm could be confined to the countries where despots and fanatics reign, then a new isolationism or confidence in the United Nations and existing international law or devotion to the principle that the oppressed peoples of sovereign nations must initiate their own revolts and reforms might in practice suffice. But no such stabilization of a status quo is possible. A dilemma must be faced: the huge obstacles to democracy in the Arab world are also the very reason that political reform is so important—indeed urgent, since the conditions that foster global terrorism and encourage tyranny are getting worse rather than better. There is a remarkable consensus in debates over democracy promotion when it comes to the diagnosis of the misery of liberalism in the Muslim and especially the Arab world. First, national wealth that derives primarily from natural resources tends to inhibit rather than enable liberalization and democratization. Second, the vanguard radicals of Islamic terrorism, as the profile of the September 11 highjackers and the masterminds of al Qaeda suggests, tend to come from those well-educated sectors of Arab society whose ambitions and aspirations are thwarted by restrictive economies and noninclusive political structures. Third, the youth-heavy demographic yields a large pool of recruits susceptible to demagogues and radical groups; according to estimates, over half the Arab world is under the age of twenty-five, and in bin Laden's native Saudi Arabia 75 percent of the population is under the age of thirty and 40 percent under fifteen, while unemployment runs 15–30 percent among men and 95 percent among women.[17]

There is a powerful temptation to conclude from such facts that liberalization and democracy in the Arab world are unattainable. Such a conclusion can seem even more compelling in light of the fact that the bungled American occupation of Iraq has undermined the work of many NGOs devoted to democracy promotion as their initiatives, especially their funneling of money to particular oppositional groups and even parties in autocratic states, are increasingly attacked as a mere instrument of American hegemony.[18] If Americans draw the conclusion that Arab liberalization and democracy are hopeless, they will fail to seek a new, more creative leadership in advancing democracy in the Muslim world and finding effective international means of dealing with outlaw regimes. The looming dilemma of American foreign policy is already visible: the architects and enthusiasts of the war in Iraq still do not want anything to do with genuine, internationally guided nation building, while those who oppose messianic unilateralism are more wary than ever of international engagements.

Untimely Meditation

The misconceptions in democratic messianism ultimately stem from the narrow understanding of the sources of freedom that has become the commonsense of American conservatism since Reaganism. Individual freedom is associated primarily—primally, so to speak—with

---

[17]Rachel Bronson, "Rethinking Religion: The Legacy of the U.S.-Saudi Relationship," *The Washington Quarterly* 28, no. 4 (Autumn 2005), p. 129. See also Zakaria, *The Future of Freedom*, pp. 119–59.

[18]Thomas Carothers, "The Backlash Against Democracy Promotion," *Foreign Affairs* 85, no. 4 (July-August 2006).

participation in the free market. A core meaning of civic democracy, namely, government's provision for the common good, was relentlessly attacked and, more importantly, redesignated and reimagined as an infringement on individual freedom. This understanding has planted itself in the structure of feeling—or, as Tocqueville would have said, the habits, ideas, and mores—of many Americans, and not just those who strongly support the Republican Party. An insecure job, lack of health coverage, and burdensome costs of educating one's children become signs, evidence, proof, of one's freedom. A sense of "self-reliance" in Emerson's language becomes associated with the precariousness of life in society. Living paycheck to paycheck without protection against unemployment and illness is evidence of freedom. As I have heard someone who lives with such precariousness say, *That's democracy, isn't it?* It is this identification of democracy with an existence exposed to the vicissitudes of the economy that is the hallmark of Reaganism's influence and its evolution through the Gingrich Congress of the 1990s and Republican dominance since 2001. A quarter century of slogans and actions has ingrained the idea that democracy and freedom flow from social precariousness.

What then becomes of civic democratic values? How do citizens experience their belonging to the body politic and their participation in it? Therein lies the corollary reduction, for if the sources of freedom are reduced to participation in, and exposure to, the free market, then political liberty as such comes to seem something derivative and secondary. The citizen's participation in the polis is easily reduced to the right to vote—and to contribute money to candidates and parties—while civic responsibility for the common good is ultimately reduced to patriotism. It is easy to parody the cowboy capitalism embraced by Bush as by Reagan—a parody made all the easier by the white Stetsons and folksy manner affected by them both—but their populist appeal has dramatically shrunk the meaning of freedom and citizenship. When carried into foreign affairs, freedom reduced to deregulated capitalism and citizenship reduced to patriotism produce the simplistic opposition of freedom and tyranny that Reagan expressed in his brand of anticommunism and Bush in the "war on terror." And when it comes to nation building, this same perspective gives rise to the attitude that democracy is attained by elections-plus-markets.

Philosophically, these neoconservative or neoliberal views rest on a *metaphysical* claim: the free market is the absolute source and ultimate goal of human freedom. This conception is a far cry from Isaiah Berlin's negative liberty. While the economic freedoms associated with capitalist industry and enterprise are in Berlin's view a valuable feature of modern society, the free market itself is a historical development that neither God nor historical necessity authored and ordained. Rather, it is an eventuality in human history whose benefits and possibilities are worth protecting and perpetuating. Berlin's orientation to tradition in the Burkean sense of valuing whatever has proved valuable to human existence over time keeps him ever on the alert against the global claims of any single value, including liberty:

> The extent of a man's, or a people's, liberty to choose to live as they desire must be weighed against the claims of many other values, of which equality, or injustice, or happiness, or security, or public order are perhaps the most obvious examples. For this reason, it cannot be unlimited. We are rightly reminded by R. H. Tawney that the liberty of the strong, whether their strength is physical or economic, must be restrained. This maxim claims respect, not as a consequence of some *a priori* rule, whereby the respect of the liberty of one man logically entails respect for the liberty of others like him; but simply because respect for the principles of justice, or shame at gross inequality of treatment, is as basic in men as the desire for liberty.[19]

And, just as important, Berlin does not give the principle of freedom any absolute, unequivocal interpretation. He defends, rather, maintaining a wide area in which individuals can exercise *freedoms*. The freedoms that are enabled by commerce, enterprise, and markets have proved valuable in innumerable and indispensible ways, but free markets are not thereby the origin, principle, and ultimate meaning of freedom. Berlin has a conservative appreciation of liberalism, affirming at once "Burke's plea for the constant need to compensate, to reconcile, to balance," and "Mill's plea for novel 'experiments in living' with their permanent possibilty of error."[20] At once conservative and liberal, his

---

[19]Berlin, "Two Concepts of Liberty," p. 170.
[20]Ibid.

thought is neither neoliberal, since he eschews any metaphysics of the free
market, nor neoconservative, since he does not seek to engineer society
and individual behavior in accordance with an ideology or religiously
inspired, state-enforced conception of family, morality, or sexuality.

Arendt also eschews a metaphysics of freedom, although tradition
and innovation play an altogether different role in her thought. Politics
is a realm of human innovation for Arendt, but this idea itself belongs
to a fragile, continually threatened, often interrupted tradition. Arendt
maintains that human freedom requires the wordly space of a politically
guaranteed public realm to make its appearance, *and* that the political
realm is itself owed to an innovation, a creation, a gratuitous act of
human freedom, which we trace back to the ancient Greeks. For it is
they who made—and recorded—the leap from clan to polis, from rit-
ual to performance, from strength to eloquence and threat to persuasion, from cult to public realm. This transvaluation of values was ungrounded: nothing necessitated it, predicted it, caused it. The creation
of the political realm is the event-without-a-cause that inaugurates
human freedom. Freedom is not a divine gift but rather the work of a
human miracle. Arendt's decisionism contrasts with Carl Schmitt's in
that the "decision" that inaugurates the political realm is not a violent
act of dominion of one or some over others but rather an unchartered
step from the prepolitical sociality of kin, clan, and cult to political
community. In the political community, belonging is defined by participation. Kinship relations remain important, as do religious rites, but
once the politically guaranteed public realm is created the spaces where
these other relations reign—the household and the temple—stand in
contrast to the public square. Belonging as it pertains to the political
community, the polis, the body politic, is a question of participation,
and is therefore utterly distinct from belonging to a tribe, family, or cult.

Arendt's caution and skepticism regarding the nation-state as it developed in European history centers on the threat that popular sovereignty poses to citizenship and the plurality of self-rule. She nevertheless does not have recourse to the sort of categorical distinction that
Jürgen Habermas makes between constitutional patriotism and nationalism or the one that Etienne Balibar makes between civic and ethnic
nationalism. Those conceptions seek to separate a good kind of political identity from a bad one in order to banish the bad. This is to seek

too much. The citizen's belonging to the polis transcends his or her "tribal" identity but does not abolish it; indeed, to put things in modern terms, civic identity does not abolish any of the citizens' other identitifications, be it nationality or religion, language or race, ethnicity or gender. The ungroundedness of the polis suggests a more unsettled, ongoing, undecided process of forging participation and belonging: civic life arises from and breaks with all preexisting, prepolitical relations, but it also at the same time must reinscribe them. Only so long as individuals' participation in public affairs holds its distinctive value will civic identity stand above and limit the scope of all other identifications. The ordeal of democracy, whether creating it or maintaining it, requires that individuals continually reanimate their political participation and civic identity; when that fails, the political realm is susceptible to all manner of sectarian, populist, or totalitarian permutations. Similarly, democracy never escapes the recurrent ordeal of finding effective symbols and myths of *fraternité*, a process that is likewise susceptible to a range of enlightened and unenlightened outcomes.

The idea that the political community is founded on an ungrounded act of agreement transcending all other forms of belonging and identity found powerful expression in American history in the Gettysburg Address, when Lincoln calls America a nation *conceived in Liberty* and *dedicated to the proposition that all men are created equal.* He thus affirms that what made America a nation was a shared idea of liberty and a mutual pledge of equality. Keeping up *that* temporary agreement of intentions and wills defines much of the drama of American democracy. In advancing his inspiring account of America's origins, Lincoln attributed the founding gesture of nation building to the Declaration of Independence rather than the Constitution. He did so in order to dispute the longstanding constitutional justification of slavery, and he did it in the midst of the civil war that he was fighting in the name of holding the nation together despite the breakdown of the agreement among its many wills and intentions. He also already foresaw that only a *new birth of freedom* would preserve the nation. Resonant in his every word is a sense of the creativity and fragility of democracy.

The Party of Lincoln has largely abandoned his legacy in recent decades. He held that the universal rights of every individual are more fundamental than states' rights; he held a tragic view of the responsibil-

ities of political office; and he believed that if Providence is manifest in
history, it is only ever as a warning to the nation of the consequences of
its unjust actions. The wisdom of tragedy is something Lincoln possessed as few other American politicians and statesmen have. As political thinkers, Arendt and Berlin respect the frailty of human affairs in
exact proportion to their affirmations of positive and negative freedom.
Political community rests for Arendt, let us recall once more, "upon the
unreliable and only temporary agreement of many wills and intentions," and she did not hesitate to state the sobering thought that "the
periods of being free have always been relatively short in the history of
mankind."[21] And Berlin saw in the ineluctable plurality of human ends
the permanent potential for conflict and tragedy: "If, as I believe, the
ends of men are many, and not all of them are in principle compatible
with each other, then the possibility of conflict—and of tragedy—can
never wholly be eliminated from human life, either personal or social."[22]
Max Weber, quite in keeping with Lincoln's understanding of the vocation of politics, discerned the tragic awareness at the heart of the ethic
of responsibility in contrast to the ethic of ultimate ends. The United
States has since September 11 been caught up in violent and uncertain
events under the leadership of a president whose proclamations for
spreading freedom are intoned with an utter denial of tragedy. The
United States is going to need the wisdom of tragedy if it is to rescue the
commitment to freedom from the wreckage of democratic messianism,
and it is going to need to draw far more amply on the traditions and experiences of democratic ideas if it is to rededicate itself to liberty and
self-rule, at home and abroad.

[21]See note 14; and "What Is Freedom?" p. 169.
[22]Berlin, "Two Concepts of Liberty," p. 169.

INDEX

abortion rights, 48, 173
Abrams, Elliot, 109
Abu Ghraib, 20, 26
Addington, David S., 133n
Agamben, Giorgio, 20, 53, 55–59
al Qaeda, 4, 85–86
al-Sadr, Moktada, 7
Anderson, Perry, 20, 127–30
Arafat, Yassir, 169
Arendt, Hannah, 3, 9–10, 13–16,
    56–59, 66–69, 179–80, 185–87, 198,
    200
Aristotle, 13, 69
Aron, Raymond, 159
Atlantic alliance, 1, 20. *See also* U.S.-
    European relations
Atta, Mohamed, 87
autocracy, 184
axis of evil, 40–42
Aziz, Tariq, 112

Baathism, 106
Baldwin, James, 74–75
Balibar, Etienne, 198
bare life, 58–59, 64, 100
Berlin, Isaiah, 16, 140, 184–88, 197–98,
    200
Berman, Paul, 18, 20, 103, 106–10, 115
bin Laden, Osama, 25–26, 85, 88, 91,
    168, 173
Blair, Tony, 118, 122–24
Blake, William, 69–70
Blix, Hans, 120–21
Bloch, Ernst, 101
Bradford, William, 68
Bremer, L. Paul III, 10–11

Bruckner, Pascal, 116
Bush, George H. W., 31–32, 37, 42
Bush, George W., 2–4, 11, 19, 24–50
    passim, 89, 106–7, 110, 117–18,
    170, 182–83; and death penalty in
    Texas, 26–28, 50; and messianism,
    46–47
Bush doctrine, 4–5, 28–30, 40, 119,
    123, 129–30, 163; and American
    exceptionalism, 29

capitalism, 43–44, 94, 97–102, 171–73;
    and East Asian Model, 188
Carothers, Thomas, 189
Chalabi, Ahmed, 132–33
Cheney, Dick, 25, 34, 36–37, 133, 161
Chirac, Jacques, 117–19, 122–25
Chomsky, Noam, 20, 87–94, 97
Christian fundamentalism, 43–44,
    170–71; and electoral politics, 47–49
Churchill, Winston, 8
citizenship, 70–71, 147; and belonging
    and participation, 71; ancient versus
    modern, 71
civil rights movement, 12
clash of civilizations, 35–36, 38, 173
Clinton, Bill, 31–32, 38
Cohn-Bendit, Daniel, 116
Cold War, 34–36, 65–66, 88, 184
cosmopolitanism, 21, 110, 141–43,
    148–49, 152, 181
covenant, 62–63, 67–68, 145

Daschle, Tom, 34
Debord, Guy, 99, 101

decisionism, 61, 66–68, 198
Declaration of Independence, 155, 199
democracy: civic, liberal, social, 16,
    140–41, 196–97; and foreign policy
    18–19, 172–73; and war, 1–3, 29–30,
    117, 137–41
democracy promotion, 189–91, 194
Democrats, 33–34, 79
Diamond, Larry, 189
Dobson, James, C., 49

ecumenism, 173–77, 179
Eliot, T. S., 76
ethic of absolute ends, 109–10; and
    Islamic terrorism, 43, 87
ethic of responsibility, 22–23, 24–25,
    45, 109–10; and sense of tragedy, 22,
    200
European Union, 123, 142–43; and con-
    stitution, 143, 161–62; as empire of
    rights, 159–65
Evans, Gareth, 153

Falwell, Jerry, 49
Fanon, Frantz, 101
fear, 8–9, 51; in Hobbes, 62–63
federation of free states, 145–46, 153,
    164–65
Feith, Douglas J., 10, 132
feminism, 13
Fischer, Joschka, 120, 125
Ford, Gerald, 27
France, 83, 113–14, 161–62
freedom: negative and positive, 21, 140,
    184–87; in foreign policy, 177–79
French Revolution, 56–57, 71, 190–91
Freud, Sigmund, 12, 160
Friedman, Milton, 189
Fukuyama, Francis, 158

Gailbraith, Peter W., 193
Garton Ash, Timothy, 160
Gbagbo, Laurent, 175–76

geo-civil war, 21, 165–69, 194–95
Germany, 83
Gettysburg Address, 199
Girondists, 192
Glennon, Michael J., 20, 127–30
Glucksmann, André, 116
Groswold, Eliza, 175
Guantánamo, 20, 26, 64
Gulf War, 4, 42, 94, 111

Habermas, Jürgen, 18–19, 20, 103–6,
    115–16, 125, 137–38, 141–57,
    160–61, 163, 198
Hamas, 169
Hardt, Michael, 20, 94–102, 159
Hassner, Pierre, 51
Hezbollah, 166–167
Hobbes, Thomas, 5–7, 37–38, 59,
    62–64, 67, 135, 144–46, 154–56,
    160; on state of nature and laws of
    nature, 63
Holocaust, 59
hubris, 9, 51
humanitarian intervention, 103, 126,
    139–40, 149–51, 153–54, 156
human rights, 56–59, 147–57; and reli-
    gion, 148
humiliation, 51, 85–86
Huntington, Samuel J., 35, 44
Hussein, Saddam, 2, 7–8, 16, 103–36
    passim
Huyssen, Andreas, 124n

international legal order, 19, 92–94, 97,
    104–6, 130–31, 141–42 . See also cos-
    mopolitanism
intifada, 44, 98, 166
Iran, 41, 167
Iran-contra scandal, 37
Iraq, 103–10; constitution, elections
    in, 1, 192; occupation of, 10–11,
    131–36, 163–64, 178–79; reconstruc-
    tion plans for, 132–35; and sanctions,
    114–15

Iraq war, 1, 16–17, 103–36 passim, 182–83; motives of, 18, 46–50
Islamism, 43–45, 85, 106, 158, 166–67, 176
isolationism, 33, 46, 180
Israeli-Palestinian conflict, 41, 44, 84, 102

Jackson, Andrew, 95
Jameson, Fredric, 84
Jefferson, Thomas, 95–96, 175
Joas, Hans, 139, 163
Johnson, Lyndon Baines, 29
judgment, 24–25, 50, 72–74, 155–56; aesthetic, 72–73, 75–77; and enlarged mentality, 73; and persuasion, 72–73; political, 60–61, 73–74, 84, 104, 116, 150–51

Kagan, Robert, 4–6, 10, 109, 137, 163–64
Kant, Immanuel, 5, 37–38, 72–73, 142, 144–46, 154–57, 160, 164–65
Kay, David, 112
Keats, John, 153
Kerry, John, 48
Kirkpatrick, Jeanne, 189
Kissinger, Henry, 18, 26
Kramer, Hilton, 76
Krepinevich, Andrew F., Jr., 192n
Kristol, William, 109
Kundera, Milan, 76
Kurtz, Stanley, 35

Lapham, Lewis, 43
Lefort, Claude, 61–62, 190n
legitimation, 60
Le Pen, Jean-Marie, 161
leviathan, 6–9, 62–63
liberal nationalism, 106
liberty. See freedom
Lincoln, Abraham, 199–200
Lukács, Georg, 101, 185

Machiavelli, Niccolò, 70, 95, 138, 160
Mamet, David, 52
Manifest Destiny, 52
Mao Zedong, 101
Marcuse, Herbert, 101
Marxism, 65–66, 98–101
Mayflower Compact, 67–69
McClellan, William, 56n
Mead, Walter Russell, 21, 171–74, 179
Melville, Herman, 6
Menchiviks, 192
mercenaries, 136
Michnik, Adam, 116
Montesquieu, 23
Mosse, George, 61
Musharraf, Pervez, 168

nation building, 37, 107, 131, 134–35, 172, 195–96
The National Security Strategy of the United States (2002), 4, 28–29, 36, 45, 123. See also Bush doctrine
NATO, 30, 82–83, 93–94
Nazism, 61–62
Negri, Antonio, 20, 43–44, 94–102, 120–21, 159
neoconservatives, 9, 108–9, 137–38, 168–69
Nicaragua, 89–91
Nietzsche, Friedrich, 69, 160
9–11 Commission, 7
Nixon, Richard M., 18, 26
North Korea, 36, 41

Packer, George, 132
Pakistan, 41, 83, 168
Paz, Octavio, 33
Pease, Donald E., Jr., 51–54, 64
Pericles, 79
plurality, 62
Pocock, J.G.A., 95
polis, 68–70, 74, 185–186; fragility of, 13–14; and civic democracy, 70, 198–99

political realm: inauguration of, 67; ungroundedness of, 66–69
Pollack, Kenneth M., 110–13, 131–34
Powell doctrine, 30–33, 40, 78–79, 134
Powell, Colin, 31, 42–43, 119–21, 133
power, 63, 95; according to Arendt, 38–39; according to Kagan, 4–6, 10, 12; enjoyment of, 27; and might, 3, 9–10, 66–67; and tragedy, 3; according to Weber, 15–16
prosletyzing, 175–77
public realm, 13–15, 72–73

Reagan, Ronald, 37, 87, 195–96
Realpolitik, 18, 138, 150
Republicans, 32, 34, 44, 47–48, 78–79, 170–71, 177
Rice, Andrew, 175–76
Rice, Condoleezza, 34, 113, 121, 133
Rieff, David, 114–15
rights of man. See human rights
Roosevelt, Franklin Delano, 8
Rove, Karl, 170–71
Roy, Arundhati, 91
Rumsfeld, Donald, 10, 30, 33, 36, 119, 134–35

Safire, William, 25–26
Said, Edward, 43–44
Sassen, Saskia, 86–87
Saudi Arabia, 168, 195
Schiller, Friedrich von, 69
Schmitt, Carl, 20, 53, 58–61, 65–69, 116, 146–47, 149, 152, 160; and Third Reich, 146
Schroeder, Gerhard, 118, 125
sectarianism, 180, 192–95
self-rule, 162, 185–87
separation and balance of powers, 55, 62, 190
September 11 attacks, 1, 24, 78–82; as act of war, 17, 81, 90
Sermon on the Mount, 45, 173
Sharon, Ariel, 44, 117

Sloterdijk, Peter, 1
Sollers, Philippe, 76
Somalia, 31–32
Sontag, Susan, 86
South Africa, 115
sovereignty, 58–69, 95–96, 116, 149; as responsibility to protect, 153–56
Spinoza, 99
state, 63–64
state of exception, 53, 55–62, 75n
Star Wars, 7, 36
Supreme Court, 27: and Hamdan v. Rumsfeld, 54–55
Syria, 167–68
suicide bombers, 86–87

Taliban, 4, 91–92
totalitarianism, 107–8, 186–87
tyranny, 64, 104, 106, 111, 154–56, 169–70; freedom-versus-, 21, 183–84

unilateralism, 18, 74–77, 82–84, 93–94, 104–5, 127–28, 131; and human rights, 147
United Nations, 30, 82, 89–90, 93–94, 112, 126–30, 143, 151; and Resolution 1441, 118–22
universalism, 71–77, 137–38, 174; and relativism, 77
USA Patriot Act, 52–53, 79
U.S. Constitution, 12, 53
U.S.-European relations, 4–5; and crisis over Iraq war, 117–25

Vattimo, Gianni, 77
Vernet, Daniel, 83
victimhood, 9, 124–25
Vietnam War, 10, 46, 94, 131; and Gulf of Tonkin resolution, 29; and Vietnam Syndrome, 32–33, 47
Villepin, Dominique de, 120–22

Wahhabism, 166, 168
Walzer, Michael, 139–40, 149–51
war on terrorism, 8, 17, 89
weapons of mass destruction (WMD),
    29, 41, 47, 111–13, 126
Weber, Max, 3, 15–16, 22, 28, 45, 109,
    138, 160, 200

Wilson, Woodrow, 172
Wolfowitz, Paul, 109, 121
Woodward, Bob, 112–13

Zakaria, Fareed, 188–89
Zapatistas, 98, 102